2D Boron: Boraphene, Borophene, Boronene

Iwao Matsuda • Kehui Wu
Editors

2D Boron: Boraphene, Borophene, Boronene

Springer

Editors
Iwao Matsuda
The Institute for Solid State Physics
The University of Tokyo
Kashiwa, Chiba, Japan

Kehui Wu
Institute of Physics
Chinese Academy of Science
Beijing, China

ISBN 978-3-030-49998-3 ISBN 978-3-030-49999-0 (eBook)
https://doi.org/10.1007/978-3-030-49999-0

This Springer imprint is published by the registered company Springer Nature Switzerland AG
The registered company address is: Gewerbestrasse 11, 6330 Cham, Switzerland

Preface

Naming of a novel material is significant to capture wide interests in societies and to develop the material science in academic and technological fields. Layers of "graphene" or "silicene" may not have been investigated such vigorously if they had been called "2D carbon (C)" or "2D silicon (Si)." 2D boron (B) is named "boraphene" by the theoretical works much earlier than the experimental discoveries of a monoatomic (monatomic) sheet of boron. Following the rule that elemental 2D materials are named as X'enes where X is the element composing the 2D material, for example, phosphorene (P) or bismuthene (Bi), the name could have been "boronene." With various candidates, the pioneering researchers gathered at an international conference and discussed the naming including the translations in Asian countries. Since "borophene" has been already used globally, we accept using this name. However, to keep our fairness in the nomenclature, we leave these nominations in the title of our book.

Boron has been an important research subject in theoretical and computational chemistry because they form a wide variety of chemical bonds with each other, making a large number of allotropes even in two dimensions. Growths of the first boron sheet were achieved by the harmonic efforts of experimentalists and theoreticians who have individually played their parts to achieve perfect B flat (Fig. 1). Up to now, a large number of boron single atom sheets, borophene, have been proposed and some of them have already been synthesized on various substrates. The growths in wide area and unique electronic structures of borophenes have attracted interests in material science and also for industrial uses. Furthermore, there have been also many atomic sheets of the boron compounds, such as h-BN and HB, that have already shown rich functionalities for future technology. The theoretical and experimental works in investigating 2D materials of elemental boron and boron compounds are expanding rapidly. For further developments in the field, it has become important to extensively review the topics by the pioneering researchers.

The book is composed of six chapters. Chapter 1 introduces a brief history of the theoretical works on boron allotropes in different dimensions: 3D (bulk), 2D (borphene), 1D (nanotube), and 0D (cluster, molecule). Recent computational works on predicting borophene are described in detail in Chap. 2. In Chaps. 3 and 4,

Fig. 1 Experimental achievements of the first boron sheet, borophene, were celebrated in mass media with humor, including music in B flat, to capture the public attentions

synthesis of borophene is described with the atomic structures and the electronic properties. In Chap. 5, frontier researches of 2D boron compounds are overviewed, featuring interaction of boron with light elements such as hydrogen, nitrogen, and oxygen. Chapter 6 reviews researches on 3D boron solids, which helps to understand the physical and chemical properties of boron.

The editing of this book was done at two sites: Institute of Physics, Chinese Academy of Science, and the Institute for Solid State Physics, the University of Tokyo. Appreciation is due to countless people who cannot all be named. Colleagues who contributed directly in clarifying the arguments presented here, collecting information, drawing figures, or critical reading of the manuscript are: Jens Kunstman, Takashi Aizawa, Motoko Yoshizawa, Yusuke Sato, Zhao Yuhao, Kayoko Sakaguchi (Kacco), and Tadashi Abukawa. The Society of Boron, Borides and Related Materials of Japan is acknowledged for offering an opportunity to discuss contents of the book during the international conference. We appreciate Boris I. Yakobson for his great contributions in developing this book.

Beijing, China — Kehui Wu

Kashiwa, Chiba, Japan — Iwao Matsuda

Contents

1 **A Historical Review of Theoretical Boron Allotropes in Various Dimensions** 1
Nevill Gonzalez Szwacki and Iwao Matsuda

2 **Borophenes: Insights and Predictions From Computational Analyses** 27
Naiwrit Karmodak, Eluvathingal D. Jemmis, and Boris I. Yakobson

3 **Synthesis of Borophene** 51
Baojie Feng, Lan Chen, and Kehui Wu

4 **Electronic Structures of Borophene** 73
Baojie Feng, Osamu Sugino, and Kehui Wu

5 **Chemically Modified Borophene** 89
Takahiro Kondo and Iwao Matsuda

6 **Physical and Chemical Properties of Boron Solids** 121
Ryosei Uno and Kaoru Kimura

Index 159

Contributors

Editors

Iwao Matsuda The Institute for Solid State Physics, The University of Tokyo, Kashiwa, Chiba, Japan

Kehui Wu Institute of Physics, Chinese Academy of Science, Beijing, China

Information on the Authors

Lan Chen Institute of Physics, Chinese Academy of Science, Beijing, China

Baojie Feng Institute of Physics, Chinese Academy of Science, Beijing, China

Eluvathingal D. Jemmis IPC Department, Indian Institute of Science, Bangalore, Karnataka, India

Naiwrit Karmodak Department of Physics, University of North Texas, Denton, TX, USA

Kaoru Kimura Graduate School of Frontier Sceinces, The University of Tokyo, Kashiwa, Chiba, Japan

Takahiro Kondo Faculty of Pure and Applied Sciences, University of Tsukuba, Tsukuba, Ibaraki, Japan

Osamu Sugino The Institute for Solid State Physics, The University of Tokyo, Kashiwa, Chiba, Japan

Nevill Gonzalez Szwacki Faculty of Physics, University of Warsaw, Warszawa, Poland

Ryosei Uno The Institute of Natural Sciences, College of Humanities and Sciences, Nihon University, Tokyo, Japan

Boris I. Yakobson Department of Materials Science and NanoEngineering and Department of Chemistry, Rice University, Houston, TX, USA

Abbreviations

ARPES	Angle-resolved photoemission spectroscopy
BD	Bond Density
BE	Binding energy
CE	Cluster expansion
CN	Coordination number
DFT	Density functional theory
DNL	Dirac nodal loop, Dirac nodal line
HD	Hole density
NEXAFS	Near-edge X-ray absorption fine structure
PSO	Particle swarm optimization
SEM	Scanning electron microscopy
STM	Scanning tunneling microscopy
TDS	Thermal desorption spectroscopy
UHV	Ultra-high vacuum
XPS	X-ray photoelectron spectroscopy

Chapter 1
A Historical Review of Theoretical Boron Allotropes in Various Dimensions

Nevill Gonzalez Szwacki and Iwao Matsuda

1.1 Introduction

Despite the great effort made by experimentalists and theoreticians to understand the structure and properties of boron allotropes, boron remains one of the less examined elements, not only in its bulk form but also in various molecular or low-dimensional forms, including clusters, layers, and nanotubes. The vast number of different atomic networks that boron atoms can form – and, linked to this, atoms with coordination numbers that vary from 2 to 6 may coexist – makes boron an extremely versatile material in terms of physical and chemical properties and, therefore, potential applications. The complex and challenging work of structure identification would not be possible today without computational modeling. The best example of the synergy between theory and experiment is a highly anticipated by computer simulations production of boron in its 2D form. Here, we provide a succinct yet comprehensive review of present understanding about the different forms that boron can adopt in various dimensions. To understand the significance of the discovery of flat boron crystals, we start our discussion with bulk boron and its several confirmed structures (of about 20 reported), all of them made of different arrangements of B_{12} icosahedral units connected directly or through additional atoms. Our review continues with planar and quasi-planar small all-boron clusters and with boron fullerenes. We also mention one-dimensional (1D) forms of boron like nanotubes. Then, we summarize 2D boron structures which, unsurprisingly, can also adopt several forms.

N. Gonzalez Szwacki (✉)
University of Warsaw, Warsaw, Poland
e-mail: gonz@fuw.edu.pl

I. Matsuda
The Institute for Solid State Physics, The University of Tokyo, Kashiwa, Chiba, Japan
e-mail: imatsuda@issp.u-tokyo.ac.jp

I. Matsuda, K. Wu (eds.), *2D Boron: Boraphene, Borophene, Boronene*,
https://doi.org/10.1007/978-3-030-49999-0_1

1.2 Bulk Boron

At least three all-boron allotropes are known, namely, the α- and β-rhombohedral and the β-tetragonal phases. In addition, α-tetragonal, γ-orthorhombic, and several other boron structures have been reported. Despite the considerable experimental and theoretical work invested in understanding the structure and properties of bulk boron allotropes, their exact number remains unknown. The reason for this is that most of the boron allotropes have a complex structure, and some of them are minimally different from others and/or are stabilized by small amounts of impurities, i.e., they are not pure boron phases. The current state of research on the different bulk boron phases and the resulting phase diagram of boron can be found in [36].

The almost regular icosahedron B_{12}, with B-atoms at the vertices, serves as the main structural motif not only of boron three-dimensional (3D) allotropes but also of all known boron-rich compounds.

An isolated regular boron icosahedron is an electron-deficient cluster, and therefore it is structurally unstable. Twelve atoms of the icosahedron provide 36 valence electrons. For a *closo*-cluster with n vertices, $2n + 2$ electrons are necessary to stabilize the cluster [45]. Therefore, 26 electrons are used for intra-icosahedral bonds. If icosahedra are arranged to form a larger cluster or bulk, another available ten electrons of an icosahedron can be involved in the formation of inter-icosahedron bonds with neighboring icosahedra. In those larger clusters, the boron atoms form not only the traditional two-center two-electron ($2c$-$2e$) bonds but also several centers that may share two electrons – nc-$2e$ bonds. This results in the complex polymorphism of boron structures in 3D. In those structures in the boron icosahedron, each atom is surrounded by five neighboring atoms and, in most cases, with one more atom from the neighboring icosahedron. For this reason, the majority of boron atoms have a coordination number of five or six.

It turns out that in real boron crystals the situation is even more complex. It was clearly demonstrated, for example, for β-rhombohedral boron, that bulk boron phases are often stabilized by the presence of point structural defects (vacancies and interstitials) and/or impurities.

First-principles studies of the role of the intrinsic defects in thermodynamic stability and electronic conduction of β-rhombohedral boron using density functional theory (DFT) have shown that the defectless β-rhombohedral boron with 105 atoms in the unit cell (β-B_{105}) is less stable than α-rhombohedral boron, and the band structure of the β-rhombohedral boron indicates valence electron deficiency and hence metallic behavior [44]. This is in contrast to the experimental results according to which the β-rhombohedral boron is the most thermodynamically stable allotrope of boron with a semiconducting nature. This inconsistency of results is mainly due to the fact that the β-B_{105} structure does not represent the real structure which has intrinsic defects that increase the number of atoms in the unit cell to a number between 105 and 108 (in some reports even higher).

In addition to the abovementioned bulk allotropes with the icosahedron as the main structural motif, several structures have been reported that are not icosahedrally related. There are some early experimental reports about the existence of a cubic allotrope of boron obtained in argon plasma experiments [40].

There are also theoretical studies of boron crystallizing in layered and quasi-layered bulk phases. In this context, it should be mentioned that the α-Ga type structure was used to study the nonmetal to superconductor transition in boron at high pressures [16]. This structure, however, has not yet been found experimentally. Recently, layered bulk boron was investigated by means of first-principles calculations in which bilayer borophenes were stacked one on top of the other and bound via van der Waals interactions [25]. This crystal, which is shown in Fig. 1.1a(left), is constructed based on the experimentally synthesized β_{12} boron sheet [9]. (Types of the boron monolayer are described in Sect. 1.5.)

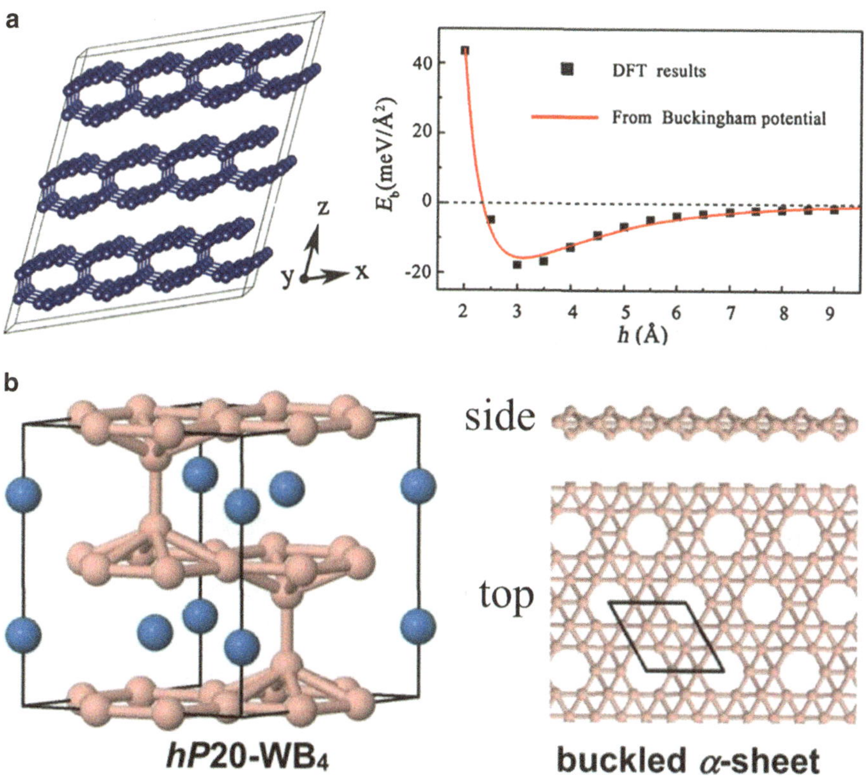

Fig. 1.1 Bulk structures based on borophene. (**a**) The structure of the bulk boron formed by bilayer β_{12} sheets via AA-stacking (left). Interlayer binding energy as a function of interlayer spacing (right) [25]. (Reprinted by permission from D. Li, Q. Tang, J. He, B. Li, G. Ding, C. Feng, H. Zhou, and G. Zhang, From Two- to Three-Dimensional van der Waals Layered Structures of Boron Crystals: An Ab Initio Study, ACS Omega 4, 8015 (2019). Copyright (2019) ACS Omega). (**b**) The $hP20$-WB_4 structure. Large and small spheres represent tungsten and boron atoms, respectively (left). Side and top views of the buckled boron α-sheet present in the $hP20$-WB_4 structure (right) [13]

In the case of boron-related solids, transition metal borides are a promising area of research for new superhard materials. Some of the most commonly studied compounds are the highest borides of tungsten, WB_{3+x}. It was recently established by means of first-principles calculations that this material is built of borophenes separated by metal atoms. Moreover, the presence of stable boron sheets is detrimental for its hardness. Very high concentrations of boron ($x \approx 1$) give rise to a soft and unstable $hP20\text{–}WB_4$ structure that can be considered to be built of quasi-planar boron α-sheets (types of the boron monolayer are described in Sect. 1.5.), separated by graphitic tungsten layers (see Fig. 1.1b).

1.3 Boron Clusters

Although the B_{12} is a common motif for all the synthesized bulk allotropes of boron, as an isolated cluster B_{12} is an unstable structure shown in Fig. 1.2a (left). A more stable form for a neutral cluster containing 12 boron atoms has an open

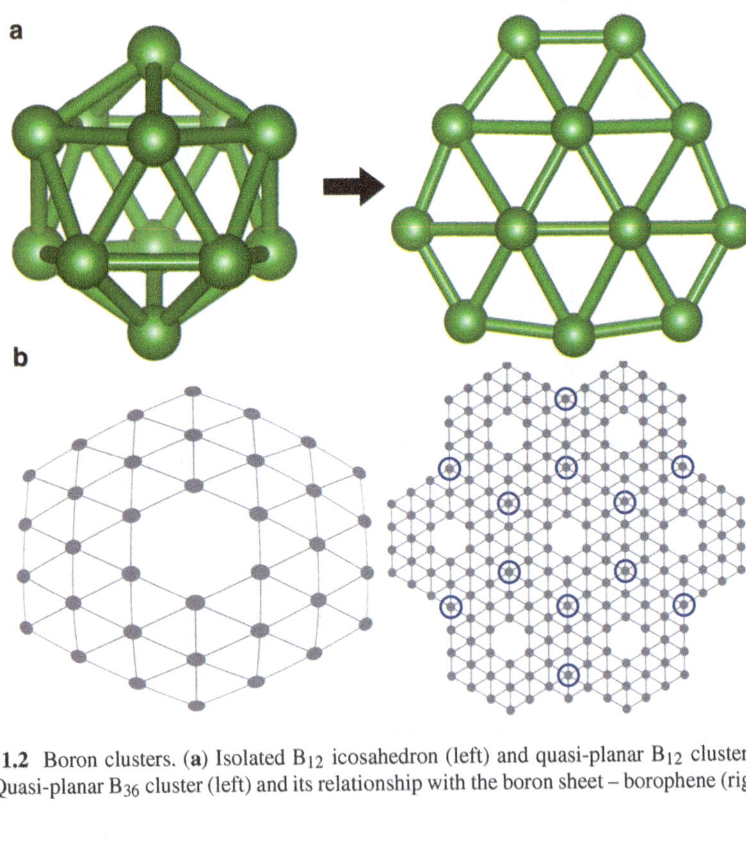

Fig. 1.2 Boron clusters. (**a**) Isolated B_{12} icosahedron (left) and quasi-planar B_{12} cluster (right). (**b**) Quasi-planar B_{36} cluster (left) and its relationship with the boron sheet – borophene (right) [31]

quasi-planar geometry, shown in Fig. 1.2a (right) [19]. This theoretical finding was the beginning of systematic computational studies of small all-boron clusters undertaken throughout the 1990s [3, 5, 18, 34].

From the experimental side, early studies – mainly involving mass spectrometry observations and chemical reactivity of small cationic boron clusters – were lacking structural or spectroscopic information [15, 22]. Extensive experimental studies of the structure and bonding of size-selected boron clusters came at the beginning of this century [17]. Combined photoelectron spectroscopy (PES) and theoretical studies have shown that anionic boron clusters, B_n^-, can be two-dimensional up to $n = 38$. Ion-mobility spectrometry (IMS) coupled with mass spectrometry and combined with theoretical calculations has established that cationic boron clusters, B_n^+, can be planar or quasi-planar up to $n = 15$.

According to bonding analyses, all the planar structures feature 2*c*-2*e* σ bonds in the edge and delocalized multicenter two-electron (*nc*-2*e*) bonds between the inner and peripheral boron atoms [17]. The delocalized σ and π bonds give rise to the concept of multi-aromaticity that is responsible for the planarity and high stability of boron clusters [50].

The history of neutral boron clusters is not straightforward. Initially, it was established that small B_n clusters favor being planar up to $n = 18$ and larger clusters prefer 3D double-ring or tubular structural motifs [20]. The 2D-to-3D structural transition established at B_{20} is reminiscent of the ring-to-cage transition at C_{20}, which forms the smallest carbon fullerene [32]. It was suggested that the B_{20} double-ring cluster may be considered as the embryo of the thinnest single-walled boron nanotube. A major breakthrough came with the synthesis of the quasi-planar B_{36} cluster shown in Fig. 1.2b (left) [31]. Its discovery reveals that the 2D-to-3D structural transition has not yet been established for neutral boron clusters. On the other hand, the quasi-planar B_{36} cluster of hexagonal shape with a central hexagonal hole can be considered as a finite fragment of an extended boron sheet shown in Fig. 1.2b (right). Therefore, it supports the viability of monolayer boron sheets with hexagonal vacancies that are common in 2D boron allotropes. Interestingly, recent studies show that the hexagonal vacancy is common for mid-sized boron clusters, while B_{26}^- is the smallest boron cluster with such a vacancy [28].

The discovery of borophenes was preceded not only by extensive experimental and theoretical work on planar and quasi-planar boron clusters but also by computational modeling of hollow boron cages. One of the first works on the topic was done employing the restricted ab initio Hartree-Fock theory [4]. In this work the studied hollow clusters were constructed with the help of the "Aufbau Principle" postulated by Boustani in 1997 [5]. The principle states that a stable boron cluster (quasi-planar, tubular, convex, and spherical) can be constructed from two basic units only: a pentagonal pyramidal B_6 unit and a hexagonal pyramidal B_7 unit. Although the proposed spherical clusters B_{12}, B_{22}, B_{32}, and B_{42} were less stable than the conjugated quasi-planar structures, this was the first attempt to model the structure and properties of hollow boron cages.

A different strategy was proposed by Szwacki *et al.* in 2007 [14] consisting of stabilization of boron clusters with similar structures to those of carbon fullerenes,

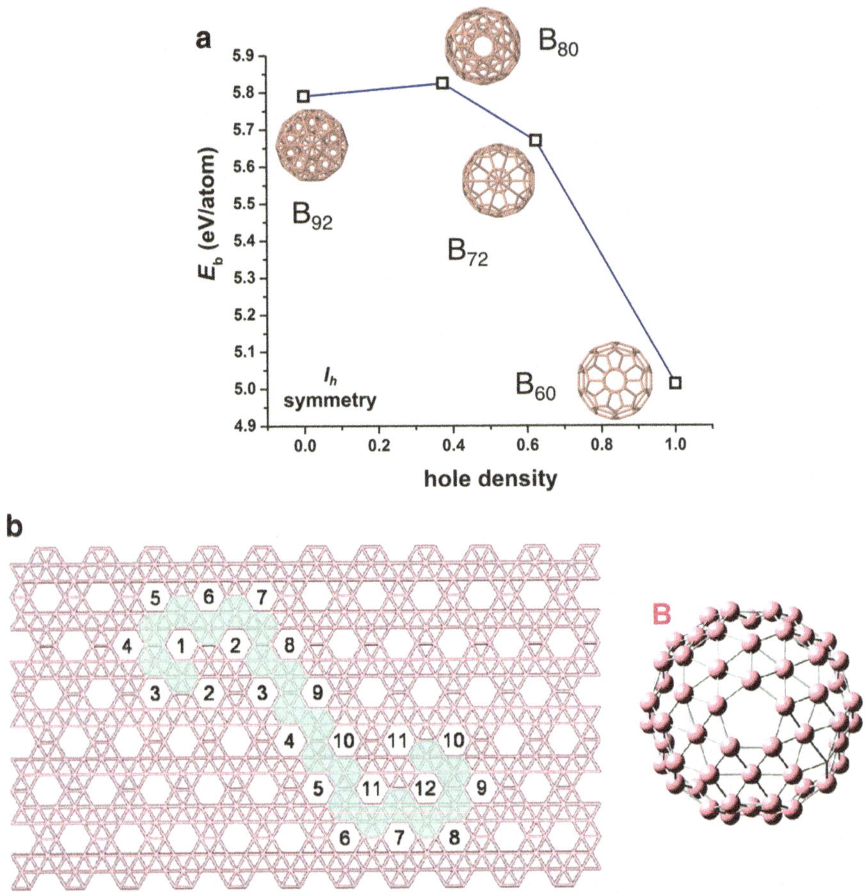

Fig. 1.3 Boron hollow spherical clusters. (**a**) Binding energy vs pentagon and hexagon hole density (number of undecorated pentagons and hexagons over 32). The corresponding cages with I_h-symmetry are shown. (**b**) Projection of the B_{80} cage (right) on the α-sheet (left). (Types of the boron monolayer are described in Sect. 1.5.) Numbered facets represent the 12 pentagons in B_{80} [35]. (By Courtesy of Boris I. Yakobson)

by decoration of the fullerene-like clusters with additional boron atoms. The B_{60} cluster analog of the I_h-C_{60} fullerene is structurally unstable in the fullerene-like cage structure. However, it can preserve the I_h-symmetry by decoration with 12, 20, and 32 additional boron atoms forming the B_{72}, B_{80}, and B_{92} hollow clusters, respectively, shown in Fig. 1.3a. The B_{72}, B_{80}, and B_{92} cages are formed by placing extra boron atoms on top of pentagons, hexagons, and pentagons + hexagons, respectively, of the B_{60} cage constrained to I_h-symmetry. Interestingly, all three decorated cages preserved the I_h-symmetry after structural optimization. As shown in Fig. 1.3a, the B_{80} cage has the largest binding energy among the I_h clusters. As shown in Fig. 1.3b, the B_{80} cage can be generated by folding the α-sheet. (Types of the boron monolayer are described in Sect. 1.5.)

1.4 One-Dimensional Boron Nanostructures

Pure 1D boron nanostructures have attracted considerable attention because of their potential applications in nanoelectronics and sensor devices. Although various 1D boron nanomaterials, such as nanowires, nanobelts, nanocones, and nanotubes, have been successfully synthesized by chemical and physical methods, this group of structures remains the least understood and seems to be the greatest challenge for researchers of boron nanomaterials [43].

In 1997, Boustani *et al.* first reported by using ab initio quantum-chemical and density functional methods, that nanotubular boron structures may consist of B_{24} clusters that can be considered as segments of extended pipes [2]. Calculations also indicated that the hexagonal pyramid B_7 cluster is the characteristic unit of convex and quasi-planar boron clusters, which may finally form boron nanosheets or boron nanotubes as shown in Fig. 1.4 [5, 6, 21]. Subsequent first-principles simulations indicated that boron nanotubes can be made of rolled-up pieces of quasi-planar

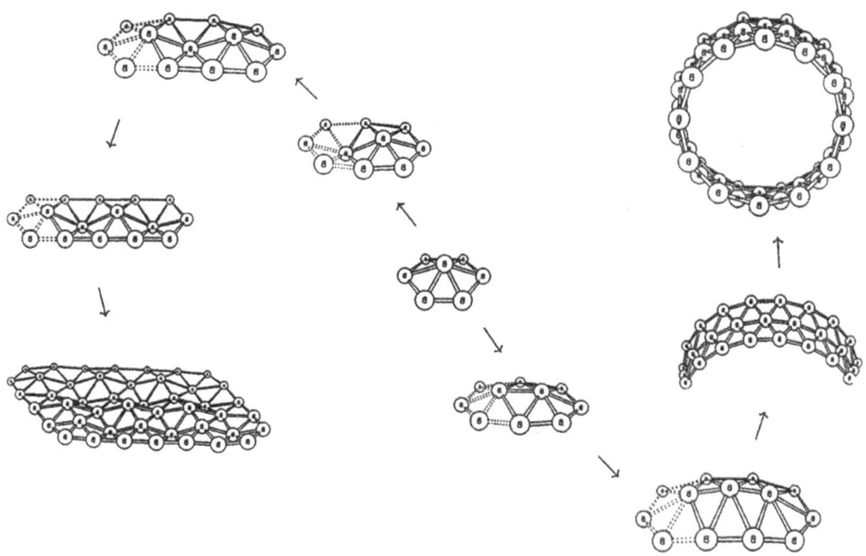

Fig. 1.4 Schematic diagram of the boron cluster growth. The key element is the hexagonal pyramid B_7 cluster. Starting from one B_7 unit, further atoms can be added to form new B_7 units, which finally leads to the formation of boron sheets and boron nanotubes [5]. (Reprinted by permission from I. Boustani, Systematic ab initio investigation of bare boron clusters: Determination of the geometry and electronic structures of B_n ($n = 2$–14), Phys. Rev. B 55, 16426 (1997). Copyright (1997) by the American Physical Society)

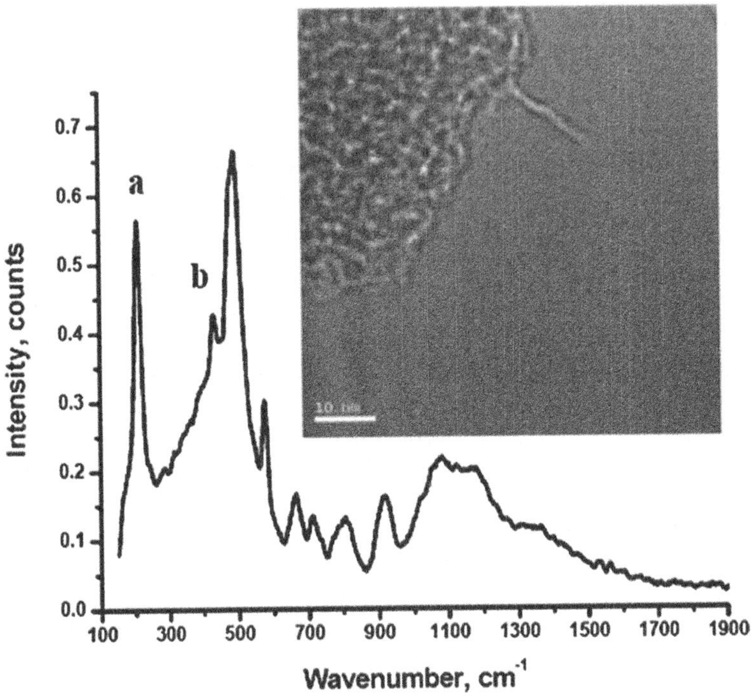

Fig. 1.5 Raman spectra recorded with "as synthesized" single-walled boron nanotubes. A single-walled tubular structure is shown in the inset micrograph [7]. (Reprinted by permission from D. Ciuparu, R. F. Klie, Y. Zhu, L. Pfefferle, Synthesis of Pure Boron Single-Wall Nanotubes, J. Phys. Chem. B 108, 13, 3967 (2004). Copyright (2004) American Chemical Society)

boron δ_6 sheets [11]. (Types of the boron monolayer are described in Sect. 1.5.) Theoretical studies have also demonstrated that boron nanotubes show metallic density of states (DOS) irrespective of their chirality [6].

Ciuparu *et al.* reported the first successful synthesis of single-walled boron nanotubes in 2004, by using a mixture of BCl_3 and H_2 as gas sources and a magnesium-substituted mesoporous silica template (Mg-MCM-41) as a catalyst (see inset of Fig. 1.5) [7]. The authors attributed the Raman peaks at about 210 cm^{-1} and 430 cm^{-1} (peaks a and b, respectively, in Fig. 1.5) to radial breathing modes of typical tubular structures. However, the structure of these nanotubes has not been confirmed as consistent with suggestions from early theoretical models.

The theoretical prediction of the new class of boron sheets, composed of triangular and hexagonal motifs, motivated new theoretical studies using the new structures as the precursors of boron nanotubes [12, 41]. In this context, Yang *et al.* studied boron nanotubes rolled-up from boron sheets (see Fig. 1.6) and demonstrated that the boron nanotubes can be either metals or semiconductors, depending on their diameter and whether they have zigzag, armchair, or chiral structure [48]. The nanotubes with diameters larger than 17 Å are all metals. Further

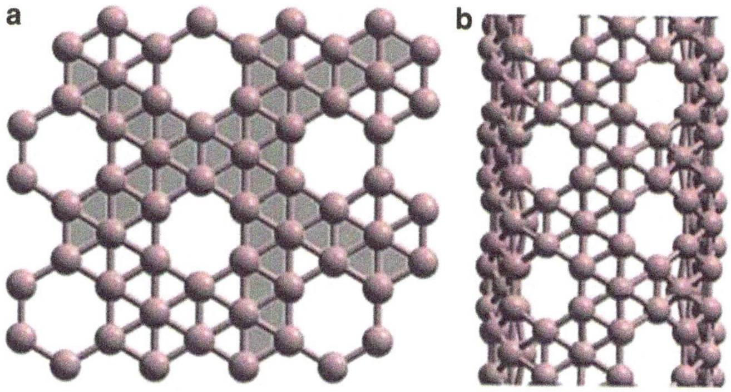

Fig. 1.6 (**a**) The structure of boron α-sheet and (**b**) the structure of a related boron nanotube [48]. (Types of the boron monolayer are described in Sect. 1.5.) (Reprinted by permission from X. Yang, Y. Ding, and J. Ni, Ab initio prediction of stable boron sheets and boron nanotubes: Structure, stability, and electronic properties. Phys. Rev. B 77, 41402 (2008). Copyright (2008) by the American Physical Society)

studies are necessary to determine whether the structure of the nanotubes is as reported in the experiment [38].

1.5 Two-Dimensional Polymorphs of Boron

A one-atom-thick boron sheet has been known to exist in crystals of metal borides composed of stacking layers of boron and metal atoms, such as MgB_2 and ZrB_2. In 2010, a surface of the $ZrB_2(0001)$ crystal was theoretically found to be composed of a graphene-like boron layer, shown in Fig. 1.7. The boron sheets, including the surface layer, in the metal borides are named "boraphene" which indicates a *graphitic boron layer* [1, 39]. Then, "borophene" was named as an extension of the 2D cluster, B_{36}, in 2014 [31]. These atomic models are drawn in Fig. 1.2b. The report has drawn researchers' attention to a new atomic layer, besides graphene, but research into 2D boron sheets was carried out much earlier and can be historically reviewed in three stages. First, researchers attempted to find the most stable free-standing boron layer in trigonal lattices. Then, it turned out that atomic defects or holes in a layer stabilize the system. In the second stage, researchers made theoretical efforts to arrange atomic holes in various 2D lattices. At the third stage, efforts have been made to investigate atomic boron sheets on various substrates both experimentally and theoretically. Then, in 2015–2016, two groups independently reported layers of borophene on the Ag(111) surfaces [9, 29]. In this section, a memory of 2D polymorphs of boron is reviewed in terms of these three stages.

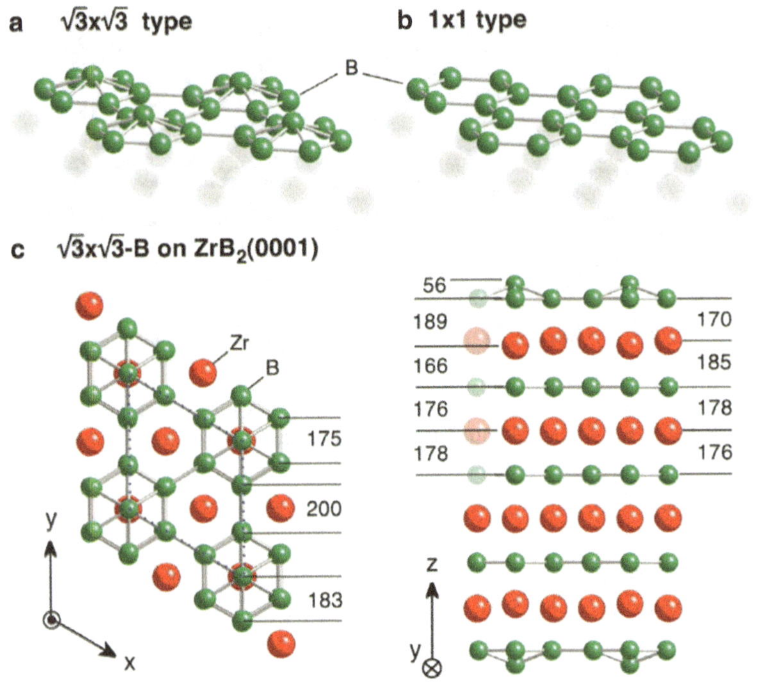

Fig. 1.7 Atomic structure models of "boraphene" on $ZrB_2(0001)$ at the B coverage of (**a**) 7/6 ML-B ($\sqrt{3} \times \sqrt{3}$ phase) and (**b**) 1 ML (1×1 phase). (**c**) Optimized atomic structure for the $\sqrt{3} \times \sqrt{3}$-B on $ZrB_2(0001)$. Atomic lengths are given in pm [39]. (Reprinted by permission from S. Suehara, T. Aizawa, T. Sasaki, Graphenelike surface boron layer: Structural phases on transition-metal diborides (0001), Phys. Rev. B 81, 085423 (2010). Copyright (2010) by the American Physical Society)

1.5.1 Free-Standing Layers

Research into bulk boron crystals has found boron atoms take a variety of bonding configurations, and there are many elemental allotropes with complicated crystal structures. This is in contrast to carbon atoms that form graphites or diamonds by the well-defined covalent bonding. Two-dimensional polymorphs of boron have been investigated based on first-principles calculations by examining various structural models and using conventional chemistry of interatomic bonding between boron atoms.

A monolayer boron was initially predicted as a sheet of the planar triangle lattice [11]. Figure 1.8 and Table 1.1 list a collection of free-standing boron sheets and their binding energies [23]. Comparing values of binding energy (BE) in the table, one finds that the graphene-like structure is less stable than the trigonal ones. The structure is called the {1212} model. Definitions of the terminology are defined in Fig. 1.8a [24]. Since there are only three valence electrons for a boron atom,

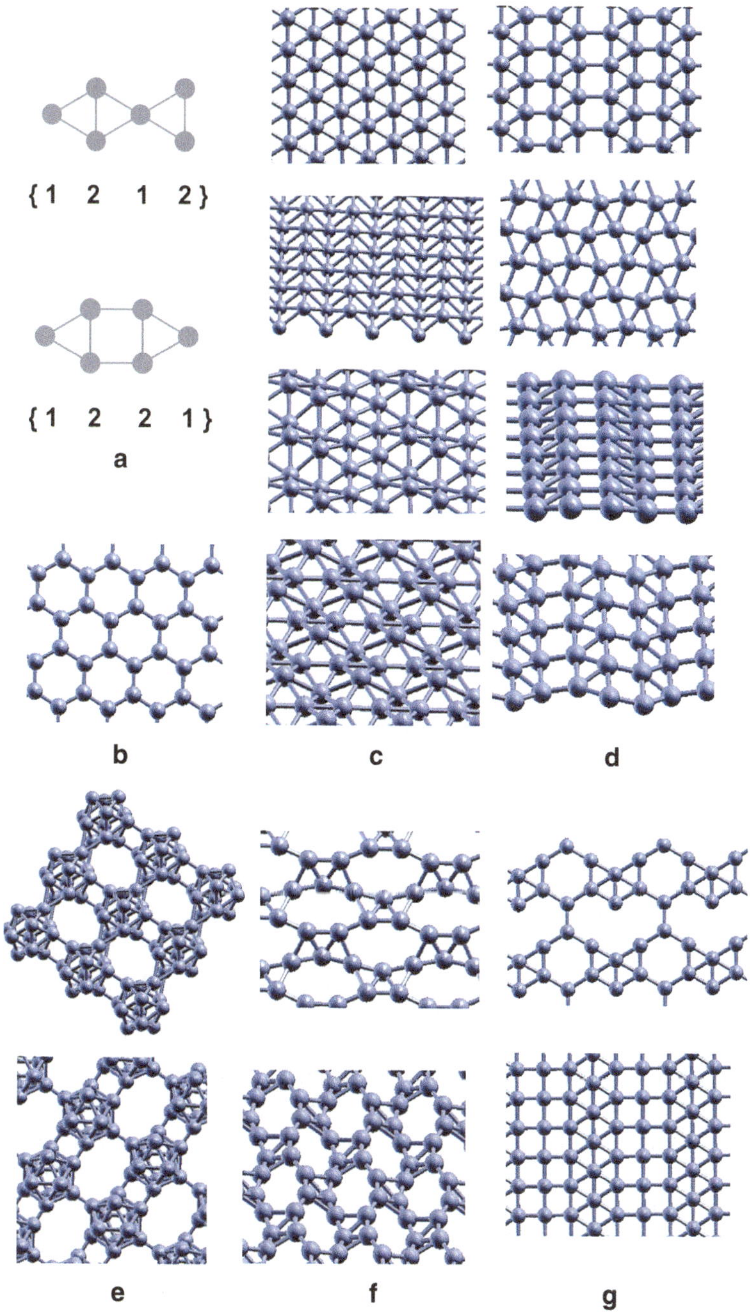

Fig. 1.8 (**a**) Definition of 1212 and 1221. Various types of boron sheets: (**b**) hexagonal graphene-like, (**c**) idealized and buckled {1212}, (**d**) reconstructed {1221}, (**e**) icosahedral, (**f**) low symmetry, (**g**) hybrid [23]. (Reprinted by permission from K. C. Lau and R. Pandey, Stability and Electronic Properties of Atomistically-Engineered 2D Boron Sheets, J. Phys. Chem. C 111, 7, 2906 (2007). Copyright (2007) American Chemical Society)

Table 1.1 A list of point symmetry, binding energy (BE, eV/atom), relative energy from the buckled {1212} (ΔE, eV/atom), and bond Lengths (RB-B, Å) for the boron sheets in Fig. 1.8 [23]. (Reprinted by permission from K. C. Lau and R. Pandey, Stability and Electronic Properties of Atomistically-Engineered 2D Boron Sheets, J. Phys. Chem. C 111, 7, 2906 (2007). Copyright (2007) American Chemical Society)

Sheet configuration	Point symmetry	BE	ΔE	R_{B-B}
buckled {1212}	C_1	5.70	0.00	1.61, 1.89
icosahedral-I	C_{1h}	5.60	0.10	1.77[a]
				1.63[b]
buckled	C_1	5.59	0.11	1.56, 1.61, 1.64
"twisted-helix"				1.78
reconstructed {1221}	C_{1h}	5.57	0.13	1.63, 1.66, 2.00
buckled {1221}	C_1	5.57	0.13	1.69, 1.76
buckled {1212}-II	C_1	5.54	0.16	1.65, 1.78
icosahedral-II	C_{1h}	5.51	0.19	1.84[a]
				1.75[b]
idealized {1212}	C_{2h}	5.48	0.22	1.71
hybrid	C_1	5.35	0.35	1.61, 1.68, 1.92
hexagonal graphene-like	C_{2v}	4.96	0.74	1.68

[a] $\langle R_{intra} \rangle$
[b] $\langle R_{inter} \rangle$

the honeycomb boron layer essentially has an electron deficiency compared to the interatomic bonding in graphene with both the σ (sp^2) and π (p_z) bonds [8]. In contrast, sixfold coordination of the boron atoms in the flat triangular lattice is not compatible with symmetry of p orbitals. Then, the buckled structure and buckled {1212} model were proposed [8, 23, 24] that allow boron atoms to form mixing σ and π bonds to reinforce the binding energy, as shown in Table 1.1. Compared to the ideal flat form of trigonal configuration {1212}, the buckled {1212} one and the buckled {1212} pair, shown in Fig. 1.8c, consist of chain-wise and pair-wise out-of-plane displacement, respectively. The reconstructed {1221} model, Fig. 1.8d, can be depicted as a network of triangle-square-triangle units, as defined in Fig. 1.8a.

K. C. Lau *et al.* analyzed sheet configurations in the {1212} category and concluded that they can be described by chemical bonding [23, 24]. The atomic coordination index, Z (i.e., number of nearest neighbors of a given atom in the 2D network), essentially determines the bonding features and stability of a given sheet. Varieties of the Z numbers appear due to a network of the combination of the localized two-center (2c) and delocalized three-center (3c) covalent bonds. The ideal case of a highly symmetrical planar {1212} sheet, $Z = 6$, is less energetically stable due to the presence of the degenerate p orbitals (Table 1.1). The electron-deficient character of the boron atom results in the nearly homogeneous delocalized bonds (Fig. 1.9-AI).

In the buckled {1212} sheet, breaking the symmetry of the planar sheet results in stability gains through the bonding configuration of $Z = 2$ and $Z = 4$ for the

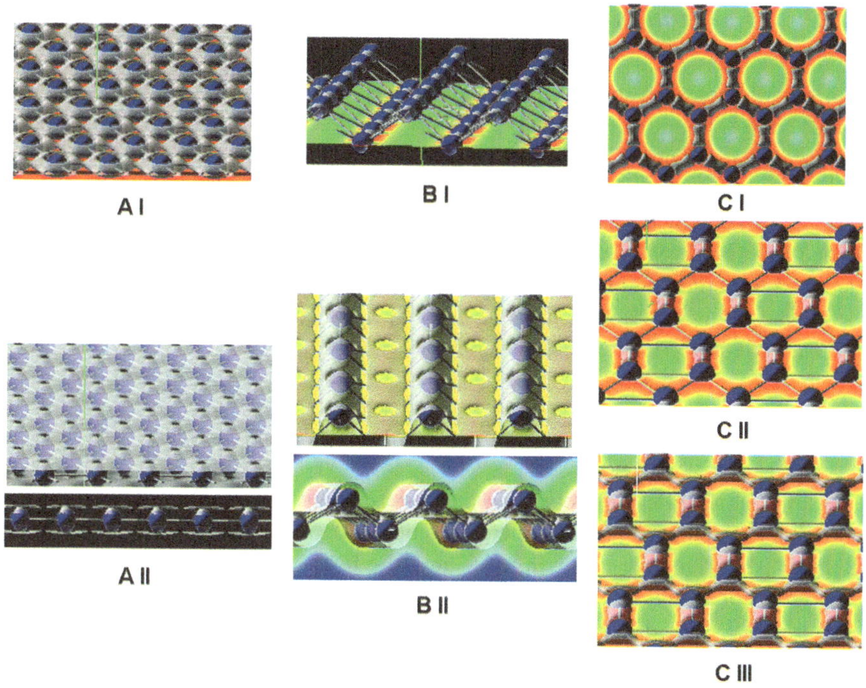

Fig. 1.9 Electron density distribution for (AI, AII) the idealized {1212}, (BI, BII) buckled {1212}, (CI) graphene-like, and (CII, CIII) reconstructed {1221} [23]. (Reprinted by permission from K. C. Lau and R. Pandey, Stability and Electronic Properties of Atomistically-Engineered 2D Boron Sheets, J. Phys. Chem. C 111, 7, 2906 (2007). Copyright (2007) American Chemical Society)

first and second nearest neighbors. As shown in Fig. 1.9-BI, the buckling forms anisotropic *s*-bonds along the infinite boron chains, breaking the degeneracy of *p* orbitals. Focusing on the boron chains, in Fig. 1.9-BII, there are "hill" and "valley" rows connected to each other by delocalized *s*-*p* bonds. In contrast to the ideal {1212} sheet, the buckled {1212} one forms a combination of localized and delocalized covalent bonds with a network of mixed 2*c* and 3*c* covalent bonds. Moving onto the reconstructed {1221} sheet, Fig. 1.8d, the layer can be viewed as a "distorted" honeycomb with new interatomic bonding. In contrast to the $Z = 3$ for the uniform honeycomb case of the graphene-like sheet, shown in Fig. 1.9-CI, the indexes are $Z = 1$ and $Z = 2$ for first and second nearest neighbors in the triangular-square-triangular network. As shown in Fig. 1.9-CII, there are directional localized *s*-bonds and delocalized 3*c* bonds along the triangular lattice networks. In the free standing {1212} boron sheets, the mixture of 2*c* and 3*c* bonds is likely essential for stabilizing the matter [24].

1.5.2 Free-Standing Trigonal Layers with Hexagon Holes

During arguments on stability of borophene between the triangular and hexagonal lattice, it was found that there is a surplus of electrons in antibonding states of flat triangular structures, while the hexagonal boron sheet is electron-deficient and requires more electrons to reduce the energy [41]. Thus, a type of borophene was designed by balancing a mixture of the triangular lattice (donor) and hexagonal lattice (acceptor). Starting with a trigonal lattice, such sheets can be structurally depicted by removing atoms and leaving hexagonal holes. Then, one can globally define the following "hexagonal hole density," η, to characterize such boron sheets with a mixture of triangles and hexagons:

$$\eta = \frac{\textit{a number of hexagonal holes}}{\textit{a number of atoms in the original triangular sheet}}.$$

Examples of relationships between η and sheets are given in Table 1.2. Figure 1.10a shows the calculated binding energy of borophene with various η values [41]. Between the triangular sheet ($\eta = 0$) and the honeycomb one ($\eta = 1/3$), there is a group of stable structures at $\eta = 0.1 \sim 0.15$ where the maximum locates at $\eta = 1/9$. The different approach of the cluster expansion method was also carried out to examine the possible borophene structure of $B_{1-x}\bigcirc_x$, where $\bigcirc$ is the hexagonal vacancy [30]. It turned out that the low-energy $B_{1-x}\bigcirc_x$ borophene structures concentrate in the vicinity of $x \approx 0.1$–0.15 ($\eta = 0.1 \sim 0.15$), as shown in Fig. 1.10b. It is intriguing that the atomic configurations examined in these studies have independent approaches but the formation energies follow the general tendency of hexagon hole density. It is of note that hexagonal borophene of the graphene-type is more unstable than the trigonal type, as described in the previous section.

Figure 1.11 shows examples of the borophene structures that were found to be stable from 9000 boron sheet structures, examined by the first-principles particle-swarm optimization global algorithm [47]. In classifying varieties of borophene, structures are labeled as α, β, χ, for example. These definitions have developed with a number of the proposed structures. The most recent was established by X. Wu *et al.* [47]. The Table 1.2 lists the names with cohesive energy (binding energy), η values, a number of boron atoms per unit cell (n), and a ratio of coordination of numbers (CN) in the structure model. The list contains the conventional names taken from literatures. Types of the boron monolayer sheets are classified as follows:

α-type: CN = 5, 6
β-type: CN = 4, 5, 6
χ-type: CN = 4, 5
ψ-type: CN = 3, 4, 5
δ-type: a single value of CN

In the case of a buckled triangle sheet ($\eta = 0$), it is CN = 6 and named as δ_6-sheet, while the honeycomb sheet ($\eta = 1/3$) has CN = 3 and is called δ_3-sheet.

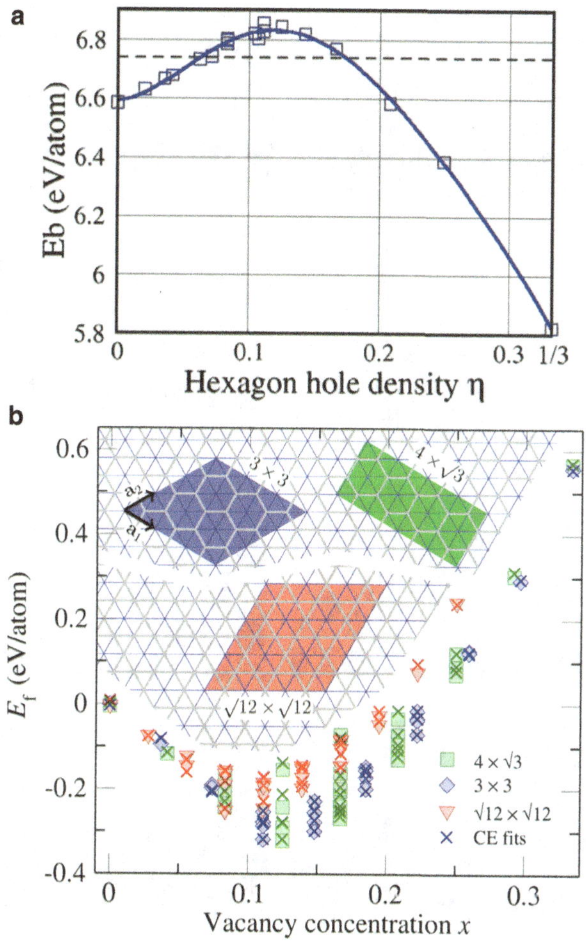

Fig. 1.10 (**a**) Dependence of binding energy, E_b, with hexagon hole density, η, for sheets that are composed of evenly distributed hexagons. E_b for the buckled triangular sheet is indicated by the dashed line [41]. (Reprinted by permission from H. Tang and S. Ismail-Beigi, Novel Precursors for Boron Nanotubes: The Competition of Two-Center and Three-Center Bonding in Boron Sheets, Phys. Rev. Lett. 99, 115501 (2007). Copyright (2007) by the American Physical Society). (**b**) Formation energies, E_f, of in the sheet system of $B_{1-x}O_x$ with various vacancy concentration x [30]. (By Courtesy of Boris I. Yakobson)

From Table 1.2, one can observe that there are various stable structures of the α, β, and χ-types, at $\eta = 1/5 \sim 1/9$, which is consistent with the results in Fig. 1.10.

It is worth mentioning previous research on stability of the α-borophene. The α-sheet has large cohesive (binding) energy, as shown in Table 1.2, and it has been regarded as the stable borophene structure. However, the calculated phonon dispersion diagram of the α-sheet shows negative phonon frequencies (Fig. 1.12) [47]. An out-of-plane bending vibration is manifested by the eigenvectors of the

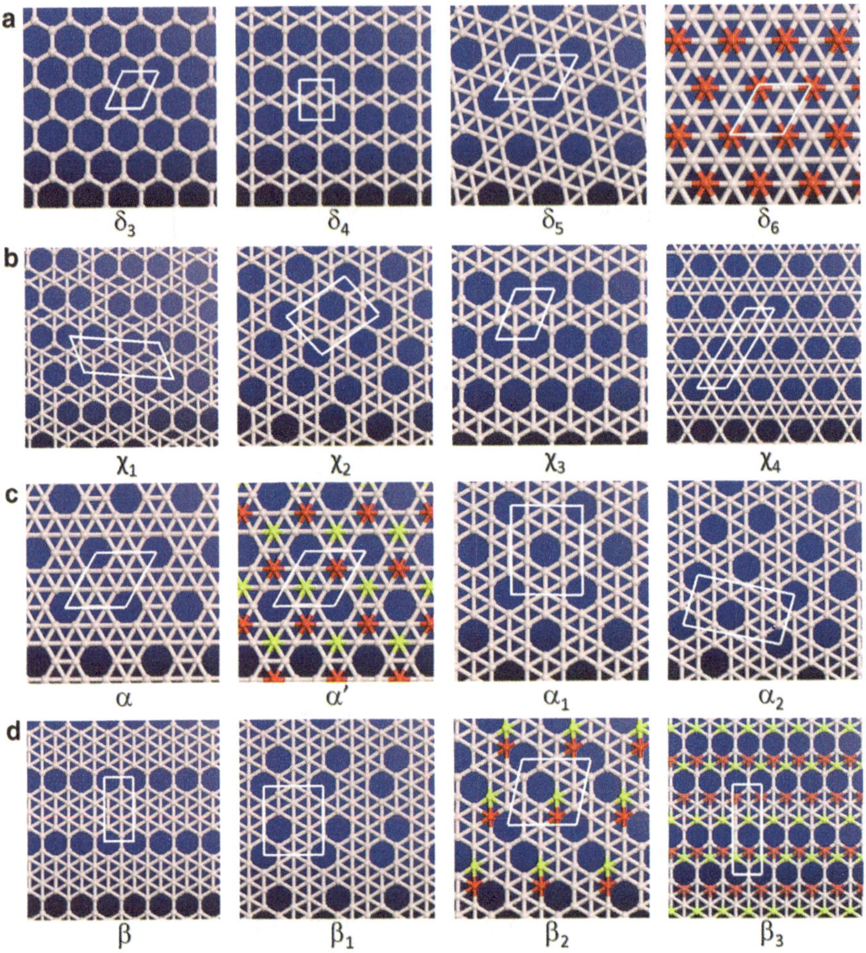

Fig. 1.11 Various structure model of the boron sheet of the (**a**) δ, (**b**) χ, (**c**) α, and (**d**) β model [47]. (Reprinted by permission from X. Wu, J. Dai, Y. Zhao, Z. Zhuo, J. Yang, X. C. Zeng, Two-Dimensional Boron Monolayer Sheets, ACS Nano 6, 7443 (2012). Copyright (2012) American Chemical Society)

negative frequency mode, followed by formation of the buckled sheet structure (α-sheet) to remove the negative frequencies. Absence of the negative phonon frequency is significant in discussing stability of matter.

Table 1.2 Cohesive energies (binding energies) of various boron sheets with difference η values and number of boron atoms in a unit cell (n) [47]. The energy is per atom. (Reprinted by permission from X. Wu, J. Dai, Y. Zhao, Z. Zhuo, J. Yang, X. C. Zeng, Two-Dimensional Boron Monolayer Sheets, ACS Nano 6, 7443 (2012). Copyright (2012) American Chemical Society)

(a)

Boron sheet	E_c(PBE)	η	n	Ratios of CN (4:5:6)
δ_3	4.877	1/3	2	
δ_4	5.384	1/4	3	
δ_5 (ref 34)	5.684	1/7	6	
δ_6 (buckled)	5.662	0	1	
χ_1	5.740	3/17	14	2:5:0
χ_2	5.740	1/6	10	1:4:0
χ_3	5.723	1/5	4	1:1:0
χ_4	5.660	1/6	10	1:4:0
χ_5	5.638		6	1:2:0
ψ	5.483	1/4	6	

(b)

Boron sheet	E_c(PBE)	E_c(PBE0)	η	n	Ratios of CN (4:5:6)
α (ref 34)	5.760	5.582	1/9	8	0:3:1
α' (buckled)	**5.762**	5.619	1/9	8	0:3:1
α_1	5.732	**5.718**	1/8	14	0:6:1
α_2	5.744	5.581	1/8	14	0:6:1
α_3	5.735	5.528	1/9	8	0:3:1
α_4	5.696		1/11	10	0:3:2
α_5 (ref 34)	5.629		1/12	11	0:6:5
β (ref 34)	5.651	5.560	1/8	7	2:2:3
β_1	5.746	**5.678**	1/8	14	1:4:2
β_2 (buckled)	5.743	5.649	1/7	12	1:4:1
β_3 (buckled)	5.737	5.639	1/6	10	2:2:1
$g_{1/8}$ (β_4) (ref 37)	5.747	5.617	1/8	21	2:5:14
$g_{2/15}$ (β_5) (ref 37)	5.740	5.612	2/15	26	2:3:8
β_6	5.734	5.600	1/6	10	2:2:1
β_7	5.736	5.573	1/7	12	1:4:1
β_8	5.724		1/9	8	1:1:2
β_9	5.719		1/6	12	1:10:1
β_{10}	5.718		1/8	7	2:2:3
β_{11}	5.712		1/6	10	3:6:1
β_{12} (ref 36; original γ-sheet)	5.712		1/6	5	2:2:1
β_{13}	5.702		1/6	10	2:2:1

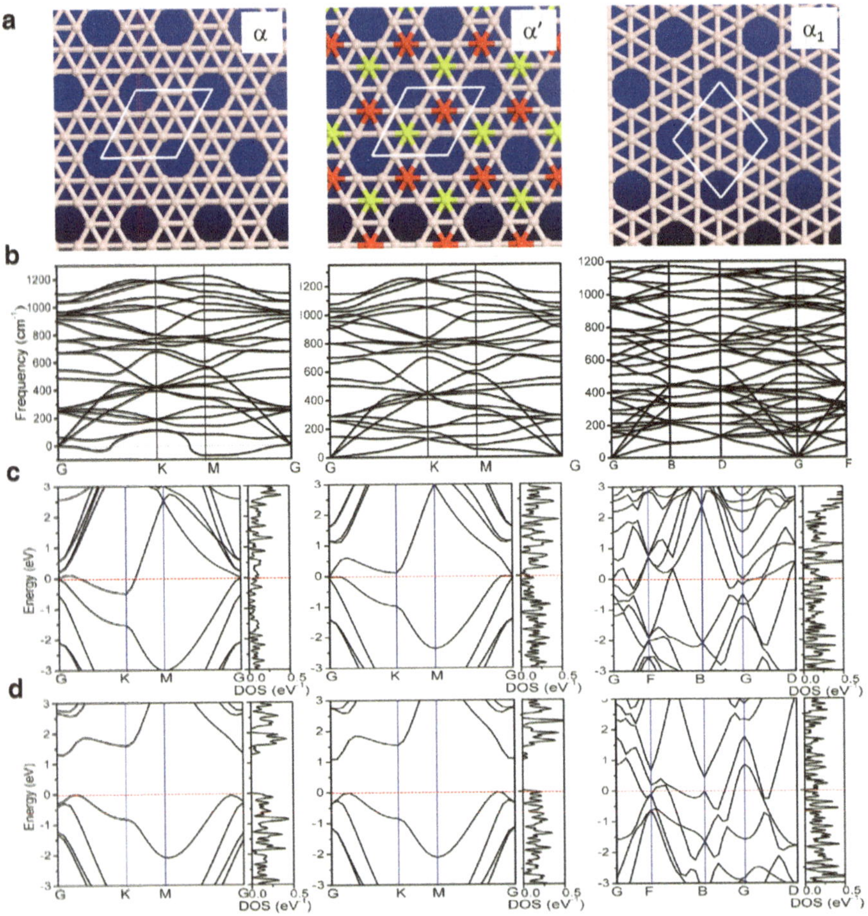

Fig. 1.12 Structure models, phonon dispersion plots, and band dispersion plots of the various α-sheets [47]. (Reprinted by permission from X. Wu, J. Dai, Y. Zhao, Z. Zhuo, J. Yang, X. C. Zeng, Two-Dimensional Boron Monolayer Sheets, ACS Nano 6, 7443 (2012). Copyright (2012) American Chemical Society)

1.5.3 Layers on Substrates

The theoretical predictions described in the previous section dealt with free-standing borophene. It has been pointed out that monolayers have two surfaces and no bulk in between them, which presents an extreme case of surface science [10]. Thus, stabilities of borophene layers may change dramatically on substrates, likely leading to novel atomic and electronic structures. The substrate effect of borophene and its synthesis have been theoretically examined. One of the landmark studies was undertaken using first-principles calculations to explore a collection of borophenes with different hole densities on metal (Au, Ag, Cu) and metal boride (MgB_2, TiB_2)

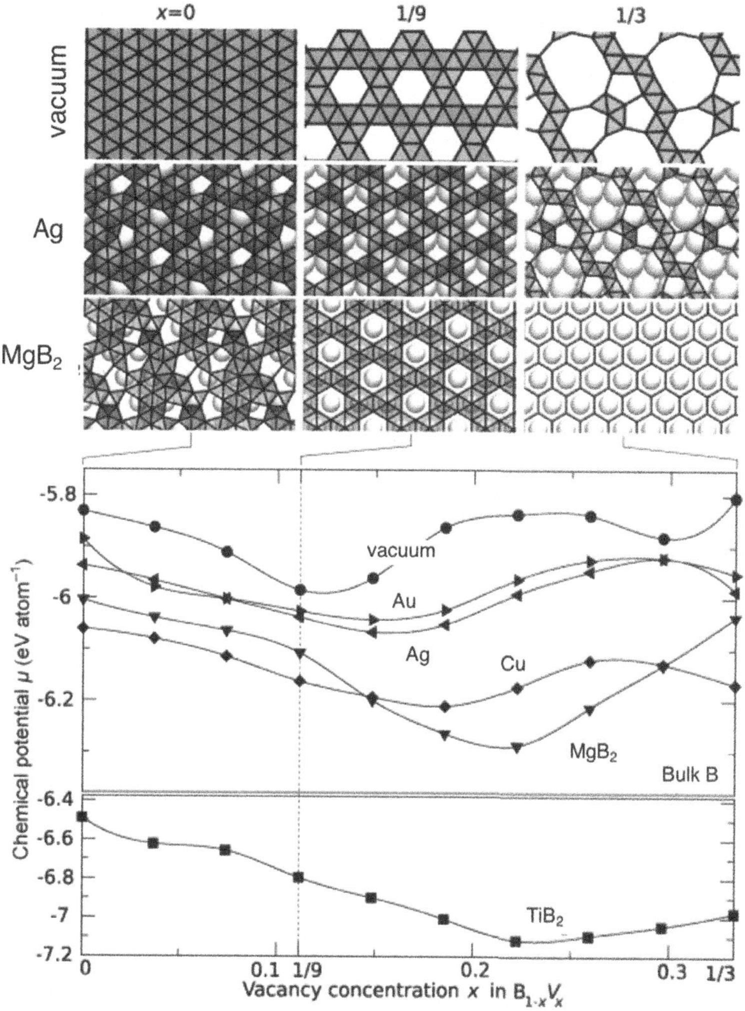

Fig. 1.13 Stabilities of 2D boron sheets with various vacancy concentrations on different substrates [27]. (By Courtesy of Boris I. Yakobson)

substrates [27]. As shown in Fig. 1.13, the calculation shows that borophene layers on the examined substrates are more stable than the free-standing ones in a vacuum. The research proposed existence of borophene when it is prepared or grown on substrates of noble metal or metal borides.

Experimentally, synthesis of borophene has been achieved on various substrates. Sheets of β_{12}- and χ_3-borophene were discovered on the Ag(111) substrate [9, 29]. The research also identified scanning tunneling microscope (STM) images of α-borophene locally [9]. A borophene phase was also synthesized on a Cu(111) surface with a single-crystal domain size of 100 μm^2 [46]. Intriguingly, a sheet of

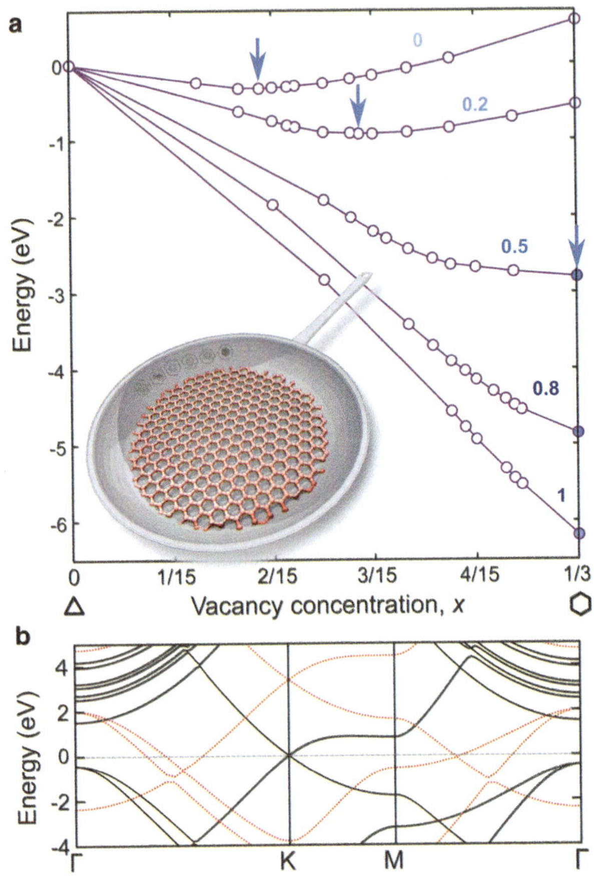

Fig. 1.14 Change of energy per boron atom of $B_{1-x}V_x$, where V is the vacancy in triangular (△) boron lattice ($x = 0$), at various doping in the range from 0 to 1 e/atom [37]. The calculation was made to describe a honeycomb borophene on a surface of aluminum. (By Courtesy of Boris I. Yakobson)

honeycomb (δ_3-) borophene was also observed on Al(111) [26]. This finding stands in sharp contrast to the fact that the free-standing honeycomb borophene is unstable, compared to other boron sheets.

Since boron atoms are electron deficient in forming 2D structures, such as a honeycomb configuration of carbon atoms, electron charging of the boron layers has been theoretically examined [37, 42]. As shown in Fig. 1.14, optimized density of vacancy increases though electron charging and honeycomb boron becomes the most stable structure when there are more than 0.5e per atom [37]. This trend can also be found in different structures of borophene [42]. These results indicate gate-voltage control of formation of borophene structures [52], opening up a new freedom when designing borophene-related electronic devices.

1.5.4 2D Boron Structures with Larger Holes

The role of vacancies in the hexagonal boron structure was studied in detail by using first-principles calculations at various levels. An early one [12, 41] was inspired by the theoretical study of the B_{80} hollow cluster. However, research on 2D boron structures was evolving rapidly, and as a consequence a polymorphism of 2D boron structures was studied by using more systematic and sophisticated numerical methods [30, 47, 49], as described above. At present, we have not only a vast number of theoretical studies available but also several allotropes of boron are already obtained experimentally on metallic surfaces [9, 26, 29] and as free-standing entities [33]. A concise summary of the theoretical and experimental studies is shown in Fig. 1.15, where we plotted the binding energy, E_b, for several borophene polymorphs as a function of the hole density. As described in the previous sections, most of the studies consider isolated boron vacancies that are commonly referred to as hexagonal holes. It is clear, however, that the number of hexagonal holes depends on the density of the 2D boron structure. For hole densities between 0 and about 0.2, the boron vacancies tend to be isolated from each other. In those structures interwoven fragments of boron double- and triple-chains (BDC and BTC, respectively) separate the holes. The most characteristic example of evenly distributed holes is probably the case of the α-sheet. However, as expected, for larger hole densities, the boron atoms tend to aggregate what leads to the formation of nonhexagonal larger holes. Examples of such porous sheets are shown in Fig. 1.16. The range of hole densities for which such structures were reported spans from 0.2 to 0.4. (Fig. 1.15) The tendency for aggregation of boron atoms is supported

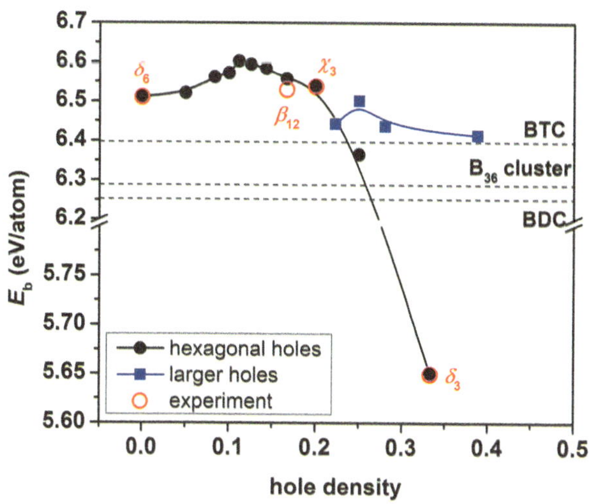

Fig. 1.15 Binding energy versus hole density for several boron 2D structures described in the text. The structures with larger holes are shown in Fig. 1.16

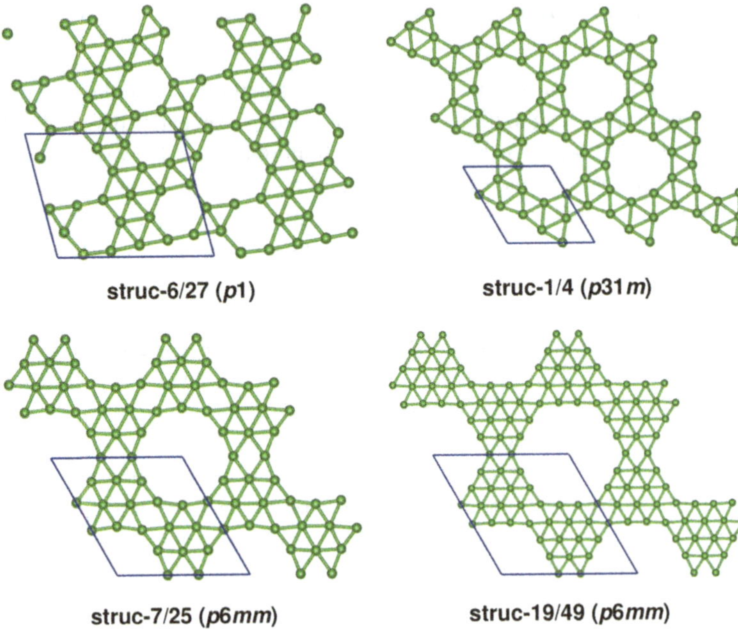

Fig. 1.16 Two-dimensional boron structures with larger holes. In parenthesis the plane symmetry groups of the structures are included. Structures struc-6/27 and struc-1/4 are illustrated from [51] and [49], respectively, whereas structures struc-7/25 and struc-19/49 are recalled from [12]

by theoretical results showing that boron atoms in 2D structures tend to be 4, 5, and 6 coordinated. The threefold coordination is highly unfavorable in terms of energy. As a consequence the formation of the honeycomb boron structure is not possible unless the structure is not stabilized by a metallic substrate, like in a recent experimental study [26]. In general, the geometry of the substrate and the electron transfer from the metallic substrate to the boron 2D structure limits the number of boron 2D allotropes that can be synthesized [51]. Interestingly, structures like the α-sheet that are not only more stable but also more isotropic than β_{12} and χ_3 have not been synthesized thus far.

References

1. T. Aizawa, S. Suehara, S. Hishita, S. Otani, Surface phonon dispersion of ZrB_2 (0001)$\sqrt{3}\times\sqrt{3}$-B. J. Phys. Condens. Matter. **20**(26), 265006 (2008) https://doi.org/10.1088/0953-8984/20/26/265006
2. I. Boustani, A. Quandt, Nanotubules of bare boron clusters: ab initio and density functional study. Europhys. Lett. **39**(5), 527–532 (1997). https://doi.org/10.1209/epl/i1997-00388-9
3. I. Boustani, Structure and stability of small boron clusters. A density functional theoretical study. Chem. Phys. Lett. **240**(1), 135–140 (1995). http://www.sciencedirect.com/science/article/pii/000926149500510B

4. I. Boustani, New convex and spherical structures of bare boron clusters. J. Solid State Chem. **133**(1), 182–189 (1997a). https://doi.org/10.1006/jssc.1997.7424
5. I. Boustani, Systematic ab initio investigation of bare boron clusters:mDetermination of the geometry and electronic structures of $B_n(n = 2 - 14)$. Phys. Rev. B **55**(24), 16426–16438 (1997b). https://doi.org/10.1103/physrevb.55.16426
6. I. Boustani, A. Quandt, E. Hernández, A. Rubio, New boron based nanostructured materials. J. Chem. Phys. **110**(6), 3176–3185 (1999). https://doi.org/10.1063/1.477976
7. D. Ciuparu, R.F. Klie, Y. Zhu, L. Pfefferle, Synthesis of pure boron single-wall nanotubes. J. Phys. Chem. B **108**(13), 3967–3969 (2004). https://doi.org/10.1021/jp049301b
8. M.H. Evans, J.D. Joannopoulos, S.T. Pantelides, Electronic and mechanical properties of planar and tubular boron structures. Phys. Rev. B **72**(4), 045434 (2005). https://doi.org/10.1103/physrevb.72.045434
9. B. Feng, J. Zhang, Q. Zhong, W. Li, S. Li, H. Li, P. Cheng, S. Meng, L. Chen, K. Wu, Experimental realization of two-dimensional boron sheets. Nat. Chem. **8**(6), 563–568 (2016). https://doi.org/10.1038/nchem.2491
10. A.K. Geim, I.V. Grigorieva, Van der waals heterostructures. Nature **499**(7459), 419–425 (2013). https://doi.org/10.1038/nature12385
11. A. Gindulyte, W.N. Lipscomb, L. Massa, Proposed boron nanotubes†. Inorg. Chem. **37**(25), 6544–6545 (1998). https://doi.org/10.1021/ic980559o
12. N. Gonzalez Szwacki, Boron fullerenes: a first-principles study. Nanoscale Res. Lett. **3**(2), 49–54 (2007). https://doi.org/10.1007/s11671-007-9113-1
13. N. Gonzalez Szwacki, The structure and hardness of the highest boride of tungsten, a borophene-based compound. Sci. Rep. **7**(1) (2017). https://doi.org/10.1038/s41598-017-04394-1
14. N. Gonzalez Szwacki, A. Sadrzadeh, B.I. Yakobson, B_{80} fullerene: an ab initio prediction of geometry, stability, and electronic structure. Phys. Rev. Lett. **98**(16), 166804 (2007). https://doi.org/10.1103/physrevlett.98.166804
15. L. Hanley, S.L. Anderson, Production and collision-induced dissociation of small boron cluster ions. J. Phys. Chem. **91**(20), 5161–5163 (1987). https://doi.org/10.1021/j100304a007
16. U. Häussermann, S.I. Simak, R. Ahuja, B. Johansson, Metal-nonmetal transition in the boron group elements. Phys. Rev. Lett. **90**(6), 065701 (2003). https://doi.org/10.1103/physrevlett.90.065701
17. T. Jian, X. Chen, S.-D. Li, A.I. Boldyrev, J. Li, L.-S. Wang, Probing the structures and bonding of size-selected boron and doped-boron clusters. Chem. Soc. Rev. **48**, 3550–3591 (2019). http://dx.doi.org/10.1039/C9CS00233B
18. H. Kato, K. Yamashita, K. Morokuma, Ab initio MO study of neutral and cationic boron clusters. Chem. Phys. Lett. **190**(3), 361–366 (1992). http://www.sciencedirect.com/science/article/pii/000926149285352B. https://doi.org/10.1016/0009-2614(92)85352-B
19. R. Kawai, J.H. Weare, Instability of the B_{12} icosahedral cluster: rearrangement to a lower energy structure. J. Chem. Phys. **95**(2), 1151–1159 (1991). https://doi.org/10.1063/1.461145
20. B. Kiran, S. Bulusu, H.-J. Zhai, S. Yoo, X.C. Zeng, L. S. Wang, Planar-to-tubular structural transition in boron clusters: B_{20} as the embryo of single-walled boron nanotubes. Proc. Natl. Acad. Sci. **102**(4), 961–964 (2005). https://www.pnas.org/content/102/4/961. https://doi.org/10.1073/pnas.0408132102
21. J. Kunstmann, A. Quandt, Constricted boron nanotubes. Chem. Phys. Lett. **402**(1–3), 21–26 (2005). https://doi.org/10.1016/j.cplett.2004.11.130
22. S.J. La Placa, P.A. Roland, J.J. Wynne, Boron clusters $(B_n, n = 2 - 52)$ produced by laser ablation of hexagonal boron nitride. Chem. Phys. Lett. **190**(3), 163–168 (1992). http://www.sciencedirect.com/science/article/pii/0009261492853196. https://doi.org/10.1016/0009-2614(92)85319-6
23. K.C. Lau, R. Pandey, Stability and electronic properties of atomistically-engineered 2d boron sheets. J. Phys. Chem. C **111**(7), 2906–2912 (2007). https://doi.org/10.1021/jp066719w
24. K.C. Lau, R. Pati, R. Pandey, A.C. Pineda, First-principles study of the stability and electronic properties of sheets and nanotubes of elemental boron. Chem. Phys. Lett. **418**(4–6), 549–554 (2006). https://doi.org/10.1016/j.cplett.2005.10.104

25. D. Li, Q. Tang, J. He, B. Li, G. Ding, C. Feng, H. Zhou, G. Zhang, From two- to three-dimensional van der Waals layered structures of boron crystals: an ab initio study. ACS Omega **4**(5), 8015–8021 (2019). https://doi.org/10.1021/acsomega.9b00534
26. W. Li, L. Kong, C. Chen, J. Gou, S. Sheng, W. Zhang, H. Li, L. Chen, P. Cheng, K. Wu, Experimental realization of honeycomb borophene. Sci. Bull. **63**(5), 282–286 (2018). https://doi.org/10.1016/j.scib.2018.02.006
27. Y. Liu, E.S. Penev, B.I. Yakobson, Probing the synthesis of two-dimensional boron by first-principles computations. Angewandte Chemie **125**(11), 3238–3241 (2013). https://doi.org/10.1002/ange.201207972
28. X.-M. Luo, T. Jian, L.-J. Cheng, W.-L. Li, Q. Chen, R. Li, H.-J. Zhai, S.-D. Li, A.I. Boldyrev, J. Li, L.-S. Wang, B_{26}^-: the smallest planar boron cluster with a hexagonal vacancy and a complicated potential landscape. Chem. Phys. Lett. **683**, 336–341 (2017). https://doi.org/10.1016/j.cplett.2016.12.051
29. A.J. Mannix, X.-F. Zhou, B. Kiraly, J.D. Wood, D. Alducin, B.D. Myers, X. Liu, B.L. Fisher, U. Santiago, J.R. Guest, M.J. Yacaman, A. Ponce, A.R. Oganov, M.C. Hersam, N.P. Guisinger, Synthesis of borophenes: anisotropic, two-dimensional boron polymorphs. Science **350**(6267), 1513–1516 (2015). https://doi.org/10.1126/science.aad1080
30. E.S. Penev, S. Bhowmick, A. Sadrzadeh, B.I. Yakobson, Polymorphism of two-dimensional boron. Nano Lett. **12**(5), 2441–2445 (2012). https://doi.org/10.1021/nl3004754
31. Z.A. Piazza, H.-S. Hu, W.-L. Li, Y.-F. Zhao, J. Li, L.-S. Wang, Planar hexagonal B_{36} as a potential basis for extended single-atom layer boron sheets. Nat. Commun. **5**(1), 3113 (2014). https://doi.org/10.1038/ncomms4113
32. H. Prinzbach, A. Weiler, P. Landenberger, F. Wahl, J. Wörth, L.T. Scott, M. Gelmont, D. Olevano, B. Issendorff, Gas-phase production and photoelectron spectroscopy of the smallest fullerene, C_{20}. Nature **407**(6800), 60–63 (2000). https://doi.org/10.1038/35024037
33. P. Ranjan, T.K. Sahu, R. Bhushan, S. SRKC Yamijala, D.J. Late, P. Kumar, A. Vinu, Freestanding borophene and its hybrids. Adv. Mater. **31**(27), 1900353 (2019). https://doi.org/10.1002/adma.201900353
34. A. Ricca, C.W. Bauschlicher, The structure and stability of B_n^+ clusters. Chem. Phys. **208**(2), 233–242 (1996). http://www.sciencedirect.com/science/article/pii/0301010496000687. https://doi.org/10.1016/0301-0104(96)00068-7
35. A. Sadrzadeh, O.V. Pupysheva, A.K. Singh, B.I. Yakobson, The boron buckyball and its precursors: an electronic structure study. J. Phys. Chem. A **112**(51), 13679–13683 (2008). https://doi.org/10.1021/jp807406x
36. K. Shirai, Phase diagram of boron crystals. Jpn. J. Appl. Phys. **56**(5S3), 05FA06 (2017). https://doi.org/10.7567/jjap.56.05fa06
37. S.N. Shirodkar, E.S. Penev, B.I. Yakobson, Honeycomb boron: alchemy on aluminum pan? Sci. Bull. **63**(5), 270–271 (2018). https://doi.org/10.1016/j.scib.2018.02.019
38. A.K. Singh, A. Sadrzadeh, B.I. Yakobson, Probing properties of boron α-tubes by ab initio calculations. Nano Lett. **8**(5), 1314–1317 (2008). https://doi.org/10.1021/nl073295o
39. S. Suehara, T. Aizawa, T. Sasaki, Graphenelike surface boron layer: structural phases on transition-metal diborides (0001). Phys. Rev. B **81**(8), 085423 (2010). https://doi.org/10.1103/physrevb.81.085423
40. D.B. Sullenger, K.D. Phipps, P.W. Seabaugh, C.R. Hudgens, D.E. Sands, J.S. Cantrell, Boron modifications produced in an induction-coupled argon plasma. Science **163**(3870), 935–937 (1969). https://science.sciencemag.org/content/163/3870/935
41. H. Tang, S. Ismail-Beigi, Novel precursors for boron nanotubes: the competition of two-center and three-center bonding in boron sheets. Phys. Rev. Lett. **99**(11), 115501 (2007). https://doi.org/10.1103/physrevlett.99.115501
42. T. Tarkowski, J.A. Majewski, N. Gonzalez Szwacki, Energy decomposition analysis of neutral and negatively charged borophenes. FlatChem **7**, 42–47 (2018). https://doi.org/10.1016/j.flatc.2017.08.004

43. J. Tian, Z. Xu, C. Shen, F. Liu, N. Xu, H.-J. Gao, One-dimensional boron nanostructures: Prediction, synthesis, characterizations, and applications. Nanoscale **2**(8), 1375 (2010). https://doi.org/10.1039/c0nr00051e
44. M.J. van Setten, M.A. Uijttewaal, G.A. de Wijs, & R.A. de Groot, Thermodynamic stability of boron: the role of defects and zero point motion. J. Am. Chem. Soc. **129**(9), 2458–2465 (2007). https://doi.org/10.1021/ja0631246. PMID: 17295480
45. K. Wade, The structural significance of the number of skeletal bonding electron-pairs in carboranes, the higher boranes and borane anions, and various transition-metal carbonyl cluster compounds. J. Chem. Soc. D 792–793 (1971). http://doi.org/10.1039/C29710000792
46. R. Wu, I.K. Drozdov, S. Eltinge, P. Zahl, S. Ismail-Beigi, I. Božović, A. Gozar, Large-area single-crystal sheets of borophene on Cu(111) surfaces. Nature Nanotechnol. **14**(1), 44–49 (2019). https://doi.org/10.1038/s41565-018-0317-6
47. X. Wu, J. Dai, Y. Zhao, Z. Zhuo, J. Yang, X.C. Zeng, Two-dimensional boron monolayer sheets. ACS Nano **6**(8), 7443–7453 (2012). https://doi.org/10.1021/nn302696v
48. X. Yang, Y. Ding, J. Ni, Ab initio prediction of stable boron sheets and boron nanotubes: structure, stability, and electronic properties. Phys. Rev. B **77**(4), 041402(R) (2008). https://doi.org/10.1103/physrevb.77.041402
49. X. Yu, L. Li, X.-W. Xu, C.-C. Tang, Prediction of two- dimensional boron sheets by particle swarm optimization algorithm. J. Phys. Chem. C **116**(37), 20075–20079 (2012). https://doi.org/10.1021/jp305545z
50. H.-J. Zhai, B. Kiran, J. Li, L.-S. Wang, Hydrocarbon analogues of boron clusters – planarity, aromaticity and antiaromaticity. Nat. Mater. **2**(12), 827–833 (2003). https://doi.org/10.1038/nmat1012
51. Z. Zhang, E.S. Penev, B.I. Yakobson, Polyphony in B flat. Nat. Chem. **8**(6), 525–527 (2016). https://doi.org/10.1038/nchem.2521
52. Z. Zhang, S.N. Shirodkar, Y. Yang, B.I. Yakobson, Gate-voltage control of borophene structure formation. Angew. Chem. Int. Ed. **56**(48), 15421–15426 (2017). https://doi.org/10.1002/anie.201705459

Chapter 2
Borophenes: Insights and Predictions From Computational Analyses

Naiwrit Karmodak, Eluvathingal D. Jemmis, and Boris I. Yakobson

2.1 Introduction

Boron allotropes have attracted interest for decades, due to their exceptional properties and large number of applications [1–3]. However, the structural complexities and multicentered bonding patterns, in most of them, cause difficulties in understanding their structure-property relationships. The two-dimensional boron monoatomic layers discovered recently on metal templates are no exception to this [4, 5]. Though most of these planar boron phases are not composed of any polyhedral fragments as that of three-dimensional allotropes, the structural controversies and possibilities of multiple polymorphs [6, 7] remind one of the early histories of 3D boron allotropes. Depending upon the crystallization techniques and the growth temperatures, the 3D boron allotropes are seen to form structures with varying number of atoms and unit cell symmetry [1–3, 8, 9]. Stability of those forms also notably varies. For the two-dimensional phases, on the other hand, slight changes in the choice of boron deposition rate, the nature of the substrate, and its temperature introduce dramatic variations in the structure [4, 5, 10–13]. This is not an artifact of experimental uncertainties, with now several laboratories around the world attempting such experiments. Is there an inherent reason for this variability?

N. Karmodak
Department of Physics, University of North Texas, Denton, TX, USA

E. D. Jemmis (✉)
IPC Department, Indian Institute of Science, Bangalore, Karnataka, India
e-mail: jemmis@iisc.ac.in

B. I. Yakobson (✉)
Department of Materials Science and NanoEngineering and Department of Chemistry, Rice University, Houston, TX, USA
e-mail: biy@rice.edu

I. Matsuda, K. Wu (eds.), *2D Boron: Boraphene, Borophene, Boronene*,
https://doi.org/10.1007/978-3-030-49999-0_2

Much before the experimental reports, several theoretical studies attempted to understand the structural possibilities of the borophene phases [14–22]. Since the discovery of graphene, the common approach to achieve the two-dimensional phases has been starting from smaller precursors and then extending along in two dimensions or exfoliating the bulk layered allotropes into its two-dimensional counterparts. However, many fragments of the three-dimensional boron allotropes are unstable in the isolated form. Moreover, apart from the polyhedral boranes, planar precursors are not known, which could be related as suitable structural motifs resulting into the borophene phases upon extension. Thus, the search for alternatives yielded some initial ideas of the structural possibilities of 2D boron. Inspired by the resemblance in chemistry of boron and carbon [23], several attempts were made to construct boron nanostructures based on the frameworks of carbon compounds. This leads to the emergence of a large number of boron structures different from the 3D boron allotropes with varying structural possibilities and properties [6, 7, 14, 24].

In this chapter, we focus on the attempts made to understand the chemistry and physics of two-dimensional boron allotropes, overviewing the theoretical works on the structural variety and stability of the different borophene phases. The discussion begins with the studies made prior to the synthesis of these phases on metals, which includes the chemistry of smaller boron clusters and the different strategies proposed to extend those clusters to 2D boron phases. This is followed by an analysis of the forces, regulating and controlling the stability and geometry of boron sheets. The structural possibilities of the recently synthesized borophene phases on metal templates are then reviewed. Finally, we summarize with a growth mechanism of the 2D phases on the metal templates focusing on the different experimental conditions, such as growth temperature, choice of boron sources, and reactivity of the metal surfaces, that play significant role in bringing the structural variations of the boron sheets.

2.2 2D Boron Clusters, Borophene's Conceptual Ancestors

Following the connection between benzenoid aromaticity and polyhedral boranes [23], several 2D boron clusters have been predicted. The idea of obtaining most probable geometries for B_n, where n varies through different values, is that with higher values of n probable structures of 2D boron phases would be obtained [14, 24, 25]. Structural preferences of aromatic hydrocarbons are often altered by addition or removal of electrons [26, 27]. This is a familiar situation known for benzene (C_6H_6) and cyclopentadienyl anion ($C_5H_5^-$), where removal of two electrons from the valence shell transforms the planar geometries into pentagonal pyramid and square pyramid structures for $C_6H_6^{+2}$ and $C_5H_5^+$, respectively. In the polyhedral boranes, each boron atom contributes two of its valence electrons toward skeletal bonding and one electron is used for exohedral B-H bond formation. Thus, based on the idea that the removal of H atoms from the polyhedral boranes would provide excess electrons and favor similar structural transformations to planar forms, attempts for stabilizing

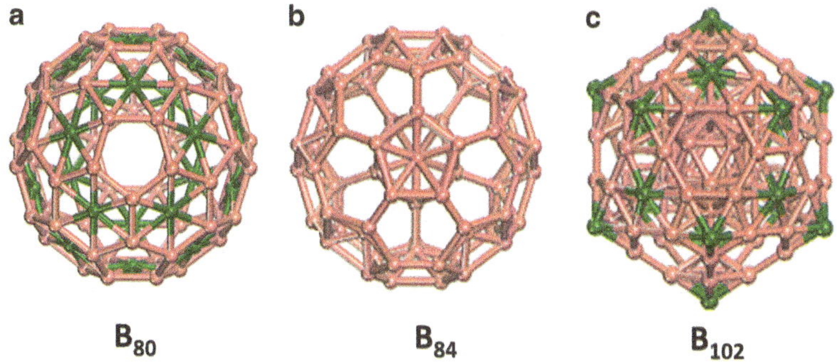

Fig. 2.1 The hollow and stuffed boron fullerenes reprinted from ref. [36] (**a**) The B_{80} cluster obtained from B_{60} by adding twenty more B atoms at the hexagonal center. (**b**) B_{84} fragment of β-boron allotrope. (**c**) The B_{102} fullerene obtained from the B_{84} by adding the 18 extra B atoms, to satisfy the required number of electrons, upon separating the fragment from the 3D unit cell

the boron clusters without the attached H atoms initiated [28–32]. A systematic theoretical study by Boustani et al. reported a number of such boron clusters for the first time [14, 24, 25]. The geometries are formed of triangular networks of boron atoms. Most of the clusters prefer a cyclic quasi-planar arrangement, with the peripheral boron rings attached with one or more number of boron atoms in the center.

Meanwhile, a stable cage cluster with 80 boron atoms, starting from the structure of C_{60} Buckminster fullerene, was predicted as the prototype for the boron fullerenes by Yakobson et al. (shown in Fig. 2.1a) [33–35]. A simple substitution of all the C atoms in C_{60} with B atoms would result in a B_{60} cluster, electron-deficient and unstable. The required 60 electrons could be provided by addition of 20 more boron atoms at the centers of the 20 hexagonal rings (green spheres in Fig. 2.1a), such that each boron atom would give its three valence electrons toward the skeletal bonding. This yields a total 80 boron atoms cluster, composed of pentagonal and triangular rings. Though a theoretical study later by Jemmis and coworkers showed that the stability of such fullerene is further enhanced by stuffing an icosahedral B_{12} unit inside the boron cage-spheres (Fig. 2.1b–c) [36, 37], the B_{80} structure was indeed the first demonstration that boron nanostructures with only triangular rings could not be stable, while a more complex pattern can.

In order to reveal the entire structural spectrum of the boron clusters with varying sizes, several experimental techniques have been developed [38–40]. A laser ablation method designed by Wang et al. and the accompanying theoretical structure search methods provide a detailed understanding of the geometries and electronic structures of these clusters [7, 41, 42], Figure 2.2 shows the structural variations in the boron clusters as the size increases. The planar clusters are obtained as the most stable isomers only up to 19 boron atoms. Above this, the cage isomers have competitive stability with respect to the planar structures. Most of these clusters also show the emergence of boron rings (denoted as holes) with greater than three

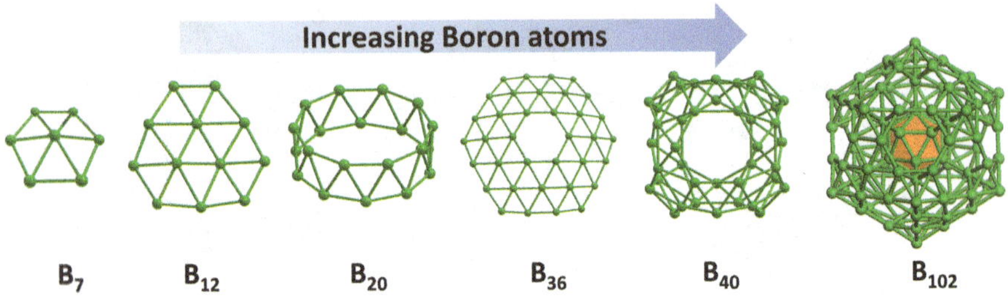

Fig. 2.2 The structural variations of the boron clusters with the increasing number of B atoms

boron atoms [7]. The structural transformations to tubular clusters are seen starting from 20 atoms and eventually become a near-spherical cluster with around 30 to 40 boron atoms [43]. The near-spherical clusters preferred stuffing with I_h-B_{12} from around a hundred or more number of boron atoms. Wang et al. introduced the term borospherenes for spherical clusters [44, 45], and from the planar cluster with a hole also the term *borophene* [46], similar to already used name *boraphene* for monoatomic boron layer in earlier Japanese study [47]. The B_{36} cluster is composed of a hexagonal ring of boron atoms in the center surrounded by 48 triangular rings. Though the search for larger boron clusters does not always lead to planar geometries, it makes it more evident that indeed the holes-vacancies are essential [16, 22, 33] for the stability of boron nanosheets. Besides, the 2D expansion of planar clusters parallels the shape of things in the realm of 2D boron allotropes. The interplay between holes (rings larger than three-membered that is atomic vacancies in the triangular lattice [22]) and the stability of 2D boron polymorphs is discussed next.

2.3 The Structural Possibilities of Borophene Phases

The structural identity of the borophene phase is subtle and has undergone several revisions since its conception decades back. Based on the structural patterns of the planar clusters, a triangular planar sheet was proposed initially out of the graphitic boron layers present in several metal diborides, especially the MgB_2 (Fig. 2.3a–c) [48]. Here, Mg atoms occupy the central positions to each hexagon (Fig. 2.3b), above and below the boron layers. With one electron less than carbon, the hexagonal unit cell with two boron atoms has a deficiency of two electrons. The electron requirement is satisfied by the charge transfer of electrons from the metal atoms, such that Mg gets a formal di-positive charge, whereas the boron becomes negative B^-. An electron added to the boron valence shell makes it isoelectronic to graphene, resulting in a stoichiometric formula $Mg^{+2}(B^-)_2$. Therefore, in order to stabilize these boron layers without any metal and obtain an all-boron sheet, it was thought that the substitution of Mg by boron would also satisfy the electron count [49, 50].

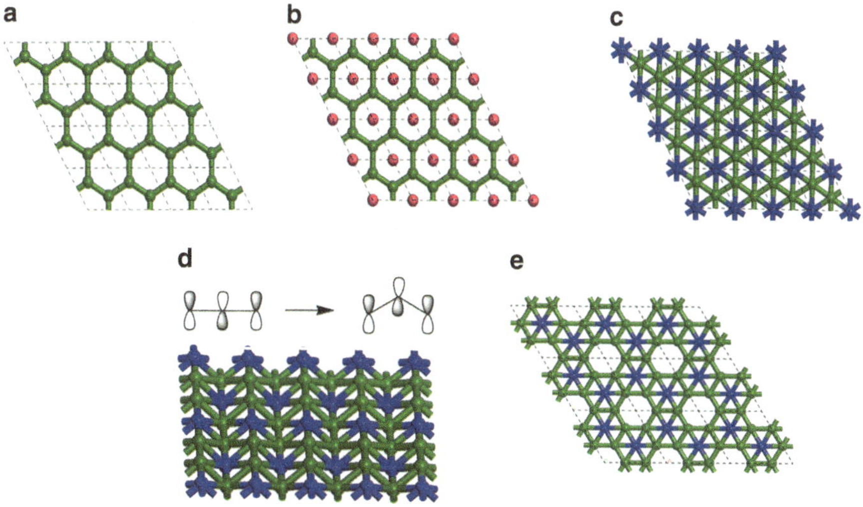

Fig. 2.3 A diagram showing the relationship between graphene and borophene. (**a**) Graphitic sheet of boron. (**b**) MgB_2 layer. (**c**) The planar triangular sheet of boron obtained from MgB_2 by substitution of Mg by B atoms. (**d**) The buckled triangular sheet and the nonplanar rearrangement of boron atoms reducing the antibonding overlap. (**e**) The α-boron sheet. (Reproduced from ref. [51])

The addition of the boron atoms introduces several differences relative to MgB_2. Firstly, the added boron atoms at the center remain in the same plane as that of the other boron atoms of the hexagonal layer, resulting in the planar triangular sheet with unit cell formula of B-B_2 (Fig. 2.3c). On the other hand, with the addition of each extra boron atom, three electrons are brought into the skeletal bonding instead of two electrons as needed per unit cell. These excess electrons would certainly occupy the antibonding bands, leading to destabilization of the sheet. The stability of the boron sheet is found to increase by an out-of-plane rearrangement of the boron atoms, since the antibonding interaction of the frontier bands is reduced [51–53]. A qualitative representation of the frontier *p* bands due to the nonplanar arrangement of the boron atoms is shown in Fig. 2.3d. Corresponding quasi-planar all-triangular sheet is also obtained from B_7 cluster following the Aufbau principle as proposed by Boustani et al. (Fig. 2.4) [24]. The hexagonal pyramid B_7 unit could extend to the curved surfaces of the fullerenes (borospherenes) and nanotubes, when the adjacent apices of the two hexagonal pyramids joined with one another by sharing a B-B bond are pointed in the same direction (pathway a in Fig. 2.4). An alternating arrangement of the apices leads to the formation of triangulated quasi-planar sheet (pathway b in Fig. 2.4).

Further, the Aufbau pyramids are quite conspicuous in cage cluster B_{80}, which also displays B-vacant pentagonal sites [33, 35], importantly suggesting that its unfolded planar form would also require a portion of B-vacant hexagons. Indeed, on the heels of this cluster proposal, Tang and Ismail-Beigi realized that in comparison

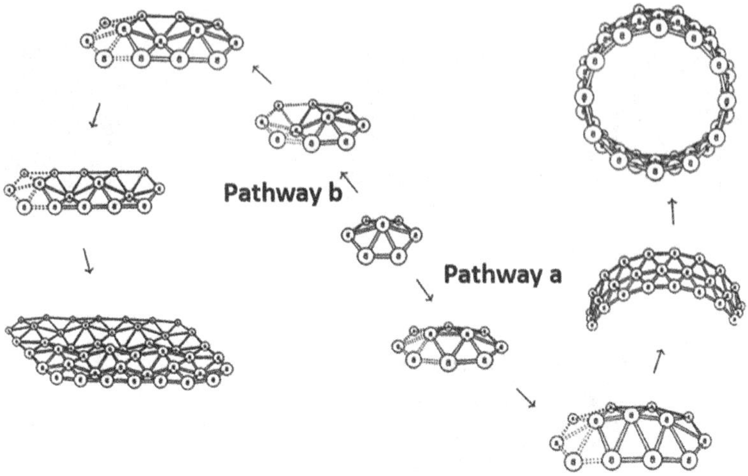

Fig. 2.4 The Aufbau principle as proposed by Boustani et al. [24]. for constructing extended boron nanostructures starting from the smaller B_7 hexagonal bipyramid cluster. The pathway a denotes the possible route to obtain the tubular structures, whereas pathway b corresponds to the quasi-planar boron sheet. (Reprinted from ref. [24] Reprinted by permission from I. Boustani, Systematic ab initio investigation of bare boron clusters: Determination of the geometry and electronic structures of B_n ($n = 2$–14), Phys. Rev. B 55, 16,426 (1997). Copyright (1997) by the American Physical Society)

to the buckling, the planar triangular sheet is better stabilized by the formation of hexagonal holes [16]. The planar triangular sheet (B-B_2) has one excess electron in each unit cell of three boron atoms. Therefore, the removal of one-thirds of the boron atoms per unit cell would reduce the excess electron density and result an electron count similar to graphene or the MgB_2 [42, 51]. In general, this could be achieved by the removal of one boron atom from every supercell of nine boron atoms (B-$B_2)_3$ (or B_3-B_6), denoted as the α-boron sheet [16]. The sheet thus obtained would be isomorphic to the unfolded B_{80} fullerene structure [33, 35]. The significance of holes becomes obvious and gives rise to the concept of the hexagonal hole density (*HD*), which is defined as [16]:

$$Hole\ density(HD) = \frac{number\ of\ hexagonal\ holes}{total\ boron\ atoms\ in\ the\ original\ triangular\ sheet\ without\ holes}$$

The triangular planar and buckled boron sheets would have $HD = 0$, whereas the α-boron sheet should have $HD = 1/9$. A number of 2D boron phases are obtained by varying the number of hexagonal holes and their arrangements within the unit cell [17, 18, 22]. The structural search methods such as cluster expansion (CE) and particle swarm optimization (PSO) are employed to reveal all possibilities of boron phases with different unit cell formula and HD. The boron sheet is assumed to be a pseudo-alloy composed of boron atoms and hexagonal holes, denoted as $B_{1-\nu}[]_{\nu}$ [19]. Here ν is the hexagonal hole (vacancy, v) density, which is varied through different values to get the optimum choice. The binding energies per atom are calculated using the first-principles approach to compare the stability of these boron sheets. Though highest stability is obtained for the sheets with HD around 1/9, several structural polymorphs are seen for each hole density distribution [19, 22]. A number of semantic classifications are made to name these phases based on the hexagonal HD per unit cell, the coordination numbers of the B atoms, and the structural arrangements. Figure 2.5 shows the structural possibilities of a few boron sheets with varying HD; the corresponding relative stabilities in vacuum, with respect to α-boron sheet, calculated using density functional theory (DFT) are given in Table 2.1, also providing the different names used in the literature to denote these sheets. Among the different nomenclatures, the one proposed based on the connectivity of the boron atoms by Wu. et al. is used more often to denote the borophene phases [17]. In the discussion here, we follow this nomenclature. As per this classification, the monolayer phases are denoted with Greek alphabets such as α, β, and so on, representing the different coordination numbers (CN) of the boron atoms, followed by arbitrary roman numerals as the subscript. The sheets with coordination numbers (CN) of 5 and 6 are named as the α-phases. The β-sheets are those having CN varying from 4 to 6, whereas the sheets with CN = 4 and 5 constitute the χ-type. The ψ-type are the sheets with CN = 3, 4 and 5. The sheets

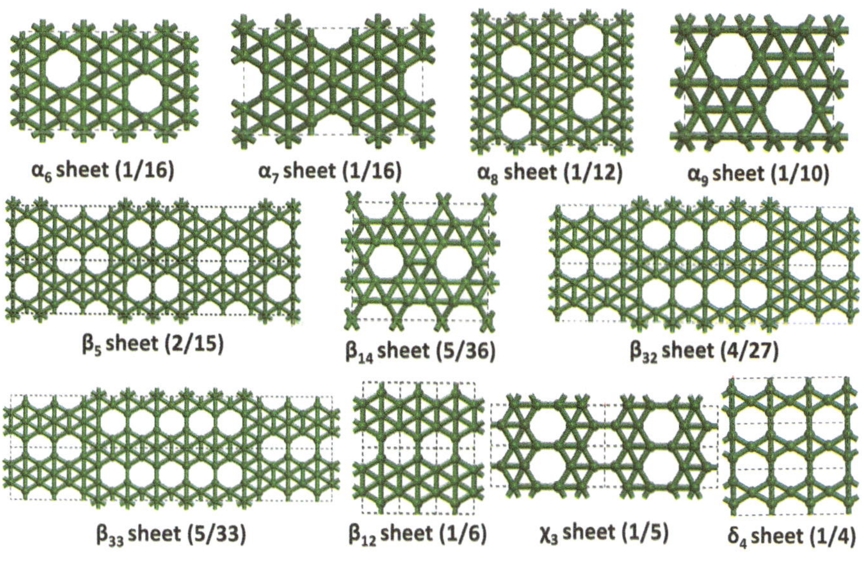

Fig. 2.5 The monolayer borophene phases with varying holes densities, HD. The boron phases are denoted using the nomenclature of Wu el. al [17]. The HD values are given within the parenthesis. (Reproduced from ref. [51])

Table 2.1 The relative binding energy (BE) of the monolayer borophene phases in meV/atom with different HD, calculated using DFT-PBE functional. The BE is reported with respect to α-boron sheet

Sheet	Notations in literature	HD	BE
δ_6'	Flat triangular sheet [16],	0	312
δ_6	δ_6 [17], buckled triangular sheet [16], Ψ_2 [54], B_Δ [55]	0	96
α_6	1/16 sheet [56]	1/16 (0.062)	82
α_7	1/16 sheet [56]	1/16 (0.062)	90
α_8	–	1/12 (0.083)	59
α_9	–	1/10 (0.100)	32
α	α' [17], struc-1/9 [18], α [16]	1/9 (0.111)	0
β_5	$\nu_{2/15}$ [19], $g_{2/15}$ [22], β_5 [17], 2/15 sheet [18], β_3 [54]	2/15 (0.133)	1
β_{14}	5/36 sheet [56]	5/36 (0.139)	59
β_{32}	–	4/27 (0.148)	50
β_{33}	–	5/33 (0.151)	28
β_{12}	β_{12} [5, 17], struc-h [18], $\nu_{1/6}$ [16], β_5 [54],$B_\square$ [55]	1/6 (0.167)	54
χ_3	$\chi\Diamond_3$ [5, 17], struc-1/5 [18], χ_1 [54], B [55]	1/5 (0.200)	43
δ_4	δ_4 [17]	1/4 (0.250)	313
δ_3	$\delta\bigcirc_3$ [17], Ψ_1 [54], hexagonal [16], B [55]	1/3 (0.333)	900

with the single value of the CN are named as δ-types; however, the subscripts in this case denote the value of the CN number. Therefore, the planar and the buckled triangular sheets are given the names δ_6 and δ_6', respectively, since each boron atom is connected to six other boron atoms, whereas the graphitic boron sheet is denoted as δ_3 phase.

2.4 Borophene Phases on Metal Substrates

The metal substrate-template introduces significant effects determining the stability of two-dimensional boron phases, in addition to the hexagonal hole density and out-of-plane buckling. The overlap with the metal bands would indeed cause electron transfer to the boron sheet. Therefore, one can assume initially that the sheets with higher electron deficiency should have improved stability on metal surfaces. This is also seen in a theoretical study by Zhang et al., where hexagonal boron sheet (δ_3) shows greater stability compared to both the planar triangular (δ_6) and α-sheet on metal surfaces, such as Mg(0001), Al(111), Ti(0001), Au(111), and Ag(111) [57]. However, later on Yakobson and coworkers found that for each metal surface, the stability would vary drastically depending upon the HD of the sheets and the experimental conditions [15, 20, 21]. Based upon the binding energies of the sheets, the metal surfaces are classified into two types. The first type consists of the metal diborides, such as MgB_2 and TiB_2, which are found to serve as the promising materials, where the boron terminated surface could provide boron phases

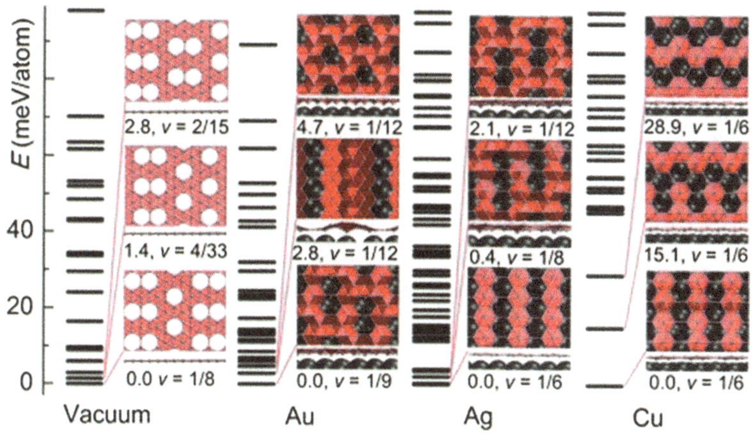

Fig. 2.6 The energy spectrum of isolated borophene phases and in the presence of Cu, Ag and Au surfaces as a function of the HD, taken from Ref. [19]. Here ν denotes the HD and the structures of the three most stable phases are shown within the insets

upon its saturation with boron atoms or evaporation of the metal atoms present underneath the surface boron layers. By tuning the experimental conditions, such as the temperature or the pressure, or even gate voltage applied to the substrate, structural preferences of the boron sheets could be varied to large extent [15, 20, 21].

On the other hand, the metal surfaces, i.e., Cu, and noble Ag and Au comprise the second category, which are suggested as the preferable substrate for the chemical vapor deposition growth of boron sheets [15, 20, 21]. Employing the structural search algorithms, the most probable polymorph of the borophene phases for each of these metal templates is determined [19]. Figure 2.6 shows the relative stability for different borophene phases with varying HD in the presence of the metal surfaces. Compared to the isolated sheets, the β_{12} sheet with 1/6 HD has the maximum stability on Cu surface, whereas the Ag surface shows preferences for several sheets with varying HD. The binding energy remains similar for a range of HD near 1/6. The Au surface has similar stability patterns as obtained for the isolated sheets. Though the sheet with 1/9 HD has the maximum stability, the relative differences in the binding energies for the other sheets are reduced compared to the values without the metal templates.

The structural possibilities of boron phases on metal templates are largely dependent upon the surface energies and work function values of the metal atoms [19, 58]. The surface energies determine whether the adsorbed boron atoms would form metal-boron alloys or proceed to the synthesis of extended 2D sheets, whereas the charge transfer possibility from the metal surfaces to the synthesized boron sheets is governed by the corresponding work function values. Having the highest surface energy, the Cu(111) surface prefers the sheets with higher HD. Though for both the Ag(111) and Au(111) surfaces, the surface energies are slightly lower, the lower work function values preferentially stabilizes the electron-deficient sheets on Ag(111) surface as well. On the other hand, the reactivity for Au(111) surface is lowest because of having lower surface energy values and higher work function. Hence, the stability trend of the boron sheets remains unchanged as for the isolated sheets. Moreover, in order to enhance the interaction with the metal substrates, the sheets undergo considerable buckling from the planarity [59].

2.5 Experimental Synthesis of Borophene

Influenced by the previous theoretical studies [15] and knowledge available from the synthetic procedures for the other elemental two-dimensional phases, the first-ever approach to synthesize monoatomic 2D boron polymorph was reported on Ag(111) surface. The two different groups succeeded in synthesizing monolayer boron phases by vapor phase deposition of boron atoms under ultrahigh vacuum (UHV) conditions. Based on the in situ scanning tunneling microscopy (STM) images, two phases are detected in one of the studies by Mannix et al. [4], denoted as striped and homogeneous phase (Fig. 2.7a–b). The lower growth temperature and higher deposition rate favor the formation of homogeneous phase, whereas with slightly higher temperature and lower deposition rate stripped phase is synthesized. The other report is by Feng et al., where the formation of two boron sheets are also reported, denoted as the S1 and S2 phases (Fig. 2.7c–d) [5]. At a temperature of around 570 K, the S1 phase is obtained, while S2 phase appears by annealing the metal template to 650 K. The STM images of both sheets show great resemblance to the phases obtained by Mannix et al. (Fig. 2.7a–d), but the structures proposed were different. The incoherence in the borophene structure interpretation eventually attracted several repeating experiments and characterizations. However, every experiment results into different consequences (Fig. 2.7e–f), such that either intermixed phases containing both the S1 and S2 phase domains are obtained [60] or completely new phases appeared [11]. This observation is generally in tune with the theory, predicting a number of 2D boron polymorphs to be formed on Ag surfaces with very similar binding energies [19, 22]

Apart from the growth temperature and the deposition rate, the structural possibilities of boron phases vary, depending upon the template surface. The deposition of boron atoms on Al (111) surface influences a honeycomb lattice structure, similar to graphene (Fig. 2.8a) [61]. On the other hand, the boron sheets formed on the Au (111) surface prefer a herringbone pattern as shown in Fig. 2.8b [10]. Two-dimensional phases are also obtained on Cu surfaces. The vapor deposition of a mixture of pure boron and boron oxide mixtures in the presence of hydrogen gas as the reducing agent led to the formation of icosahedral sheets with a thickness of around 0.8 nm[12], whereas monolayer phases are formed when only boron vapors are used at 770 K with a deposition rate of 0.05 monolayer area per minute (ML min^{-1}) [13]. The growth kinetics and the sizes of the boron sheets varied to great extent on Cu template, compared to both Ag and Au surfaces (Fig. 2.8c).

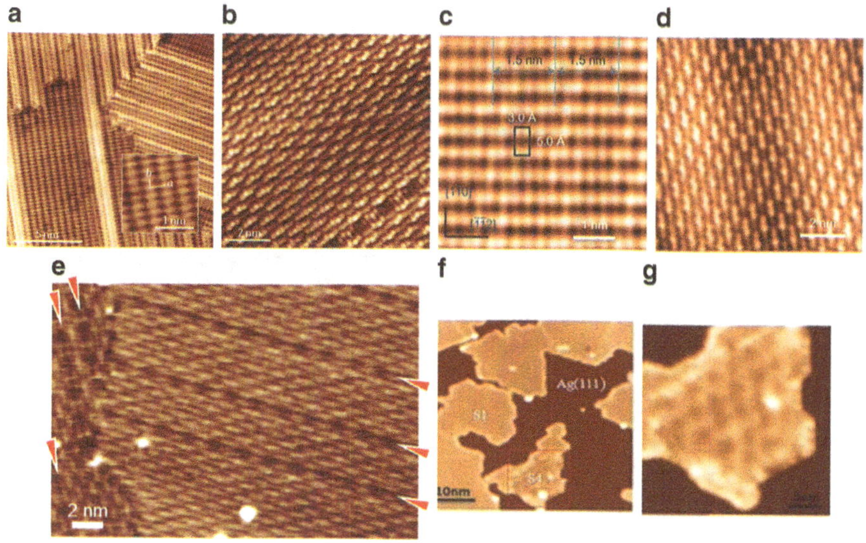

Fig. 2.7 The experimental STM images of the borophene phases obtained on Ag(111) surface reproduced from ref. [4, 5, 11]. (a) and (b) The stripped and the homogeneous phase reported by Mannix et al [4]. (Reprinted by permission from A. J. Mannix et al., Science 350, 1513 (2015). Copyright (2015) American Association for the Advancement of Science. (c) and (d) The S1 and S2 borophene phases reported in ref. by Feng et al [5]. (By courtesy of Kehui Wu). (e) The intermixed phase from S1 and S2 phases obtained by Liu et al. The red arrows denoted the separating regions between the two phases. (f) and (g) The formation of S4 phase along with the S1 phase reported in Ref. [11] (By courtesy of Kehui Wu)

The structural models of synthesized borophene phases on metal templates include the following.

2.5.1 Borophene Phase on Ag(111) Surface

S1 Phase and the Striped Phase The structural models proposed for the S1 and stripped borophene phases brought in some stimulating controversy. The STM images for both the stripped phase and S1 phase contains parallel arrangement of lighter and darker protrusions, with periodicities of 3 Å along the rows and 5 Å across the rows (Fig. 2.6a–b) [4, 5]. The linear arrangement of the lighter protrusions corresponds to the electron-rich regions, whereas the darker protrusion denotes the region with electron deficiency. The S1 phase is predicted to match with the β_{12} boron sheet with the HD of 1/6, whereas the stripped phase was correlated to the buckled triangular sheet with zero HD (denoted as δ_6' sheet). However, in later theoretical and experimental studies, it has been seen that both the phases are similar, pointing to the β_{12} boron sheet (Fig. 2.9a) [51, 54, 62], which has been

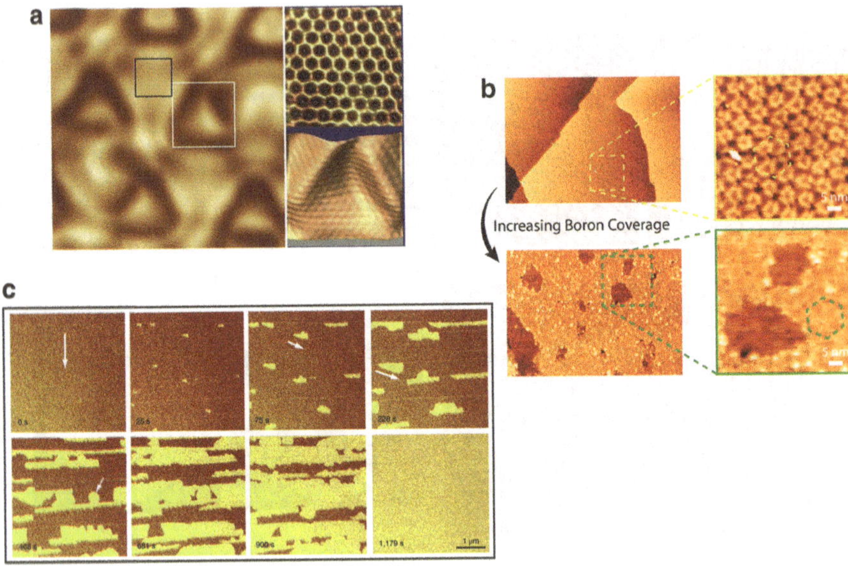

Fig. 2.8 (**a**) The hexagonal boron phase obtained on Al (111) surface [61]. (By courtesy of Kehui Wu). (**b**) The herringbone arrangement of the smaller boron islands formed by the deposition of boron atoms on Au (111) surface [10]. The smaller islands form the extended phases with higher boron coverage. (**c**) The monoatomic boron phases formed on Cu(111) surface at different time period [13]. (Reprinted from ref. [10, 13, 61] (By courtesy of A. Gozar))

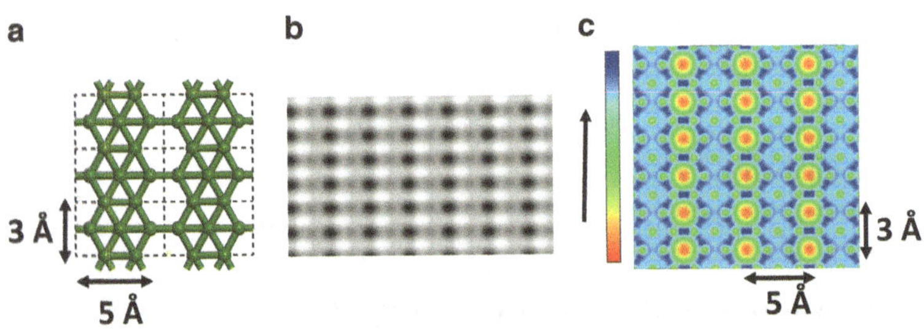

Fig. 2.9 The structure (**a**), simulated STM image (**b**), and the electron density distribution plot of β_{12} sheet (**c**). (Reproduced from ref. [51])

pointed out earlier theoretically as the most favorable phase on the Ag(111) surface by Zhang et al. [19]. The simulated STM images for the β_{12} boron sheet is shown in Fig. 2.9. The filled and vacant (holes) hexagonal units, arranged in linear fashion for β_{12} sheet, give rise to the linear patterns of lighter and darker protrusions in the simulated STM image (Fig. 2.9b–c). The filled hexagons in the β_{12} sheet have the lateral length of 3 Å, whereas the longitudinal length is around 5 Å.

S2 Borophene Phase and the Homogeneous Phase on Ag(111) Surface The S2 borophene phase has a brick-wall-type arrangement of the electron-rich and -

deficient regions, as the STM images (Fig. 2.7d) show alternate patterns for the lighter and darker protrusions [5]. Though much detailed information is not obtained for the homogeneous phase, STM image shows a similar pattern as the S2 phase, shown in Fig. 2.7b [4]. The χ_3 sheet with HD of 1/5 is initially predicted to be the precise structure for the S2 phase, despite periodicities have slight disagreement with the experimental values. The interrow periodicity is around 8.41 Å, and along the rows, it is 2.99 Å for χ_3 sheet, while the corresponding experimental values are around 15 Å and 4.3 Å, respectively. However, S2 phase could have multiple structural possibilities, depending upon its growth conditions [51].

The transformation of the S1 phase to S2 phase has been observed during the experiment upon increase in temperature [5]. This rearrangement is to readjust the electron requirement due to nonplanar undulations in the sheets with enhancement of temperature [59]. The overlap between the β_{12} sheet and the Ag surface would vary from one surface region to the other, depending upon the arrangement of the boron and the surface metal atoms. For the regions where the overlap would be greater, the HD of the sheet would increase further from 1/6 such that the greater charge transfer from the metal surface could be accommodated. On the other hand, the HD would reduce for those regions of the sheet where the connectivity between the sheet and the metal surface gets reduced. The χ_3 sheet with HD of 1/5 is one of the possibilities for the phases with enhanced HD. The other possibilities could be the β (β_5, $\beta_{32}$and β_{33}) sheets with slightly lower HD and formed of alternating filled and vacant hexagons as shown in Fig. 2.10. The relative stability of these sheets on Ag surface is slightly less compared to χ_3 sheet. Alternating filled and vacant hexagonal patterns would form a brick-wall-type arrangement for the lighter and darker protrusions in the corresponding simulated STM images (Fig. 2.10). The periodicities of β_5 sheet is 14.62 Å and 5.06 Å along and across the rows, respectively, which are almost similar as obtained for the S2 phase. The other two β-sheets have slightly higher periodic length along the rows; however, the arrangement across the rows coincides with the S2 phase.

2.5.2 *Borophene Phases Synthesized on Cu Surface*

The structural possibilities of 2D boron phases on Cu surface are largely dependent upon choice of boron sources as also evident from the recent synthetic reports [12, 13]. The boron nanostructures formed from deposition of boron oxide vapors on metal templates have been studied in greater details by Sun et al. [63]. The mechanism follows several complex steps. The initial deposition of the boron oxide vapors leads to the formation of boron oxide nanostructures on the metal surface. Later on, when these nanostructures are heated in the presence of the reducing agents, the oxygen atoms are removed as water vapors, and the nanostructures undergo rearrangement to metastable phases. During this rearrangement, the boron atoms prefer the formation of icosahedral units connected among each other. On the other

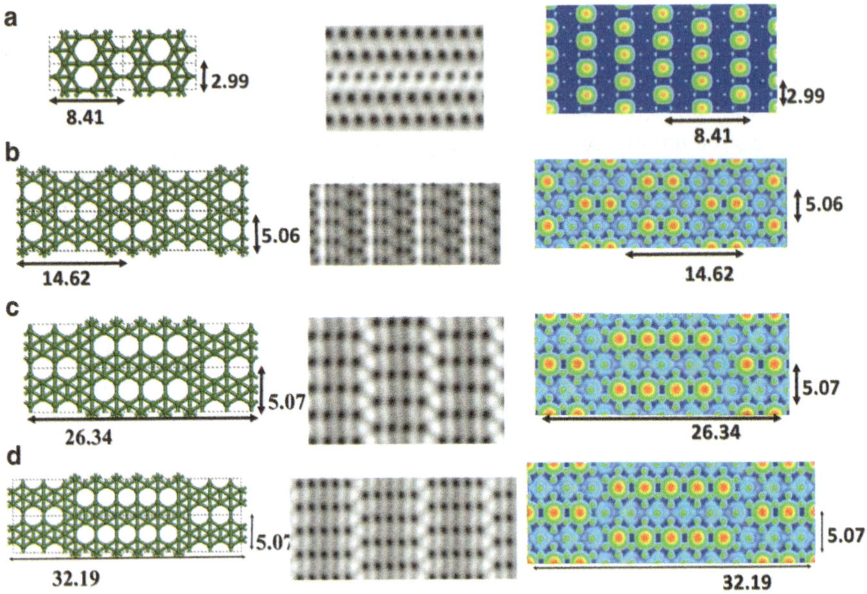

Fig. 2.10 The lattice parameter, simulated STM image, and electron density plots, shown from left to right, for (**a**) χ_3 sheet, (**b**) β_5 sheet, (**c**) β_{32} sheet, and (**d**) β_{33} sheet, respectively. (Reproduced from ref. [51])

hand, the deposition of boron atoms results into formation of monoatomic layers with larger domains. In combination of the theoretical studies and experimental STM images, the most favorable phase is found to have the HD of 1/5; however, the atomic arrangement differs slightly from the χ_3 sheet obtained on Ag(111) surface. The boron layers show corrugation along the perpendicular direction. The unit cell dimensions of the phase is around 21.84 × 15.96 Å, much larger compared to both the β_{12} and χ_3 sheets [13].

A similar scenario is obtained from the molecular dynamics study of the boron clusters and extended boron sheets on Cu(111) surface [58]. The icosahedral clusters disintegrate into planar conformations, but an extended sheet formed of Ih-B_{12} units shows considerable stability and does not undergo any structural transformations. Therefore, the icosahedral sheet is apparently metastable in nature and would rarely disintegrate into planar phases after its formation. However, the formation of the icosahedral sheet would not be possible by boron atom deposition and would need prior control on the experimental conditions.

2.5.3 *Borophene Phases Synthesized on Au Surface*

The Au surface shows differing consequences compared to both the Ag and Cu surfaces because of its inert nature [10]. The Au (111) surface undergoes significant reconstruction, depending upon the surface temperatures, while herring-bone reconstruction is observed during the initial stages with lower temperatures, which eventually transform to extended triangular networks with the increase in temperature. Due to the surface reconstruction, the deposition of the boron atoms results into different consequences as well. The lower temperature favors the finite clusters on the surface, whereas upon increasing the temperature, initially the deposited atoms get dissolved with surface Au atoms and diffuse into the bulk Au. Upon cooling to around the room temperature, the boron atoms move to the surface and segregate to smaller rhombohedral islands of boron phases ranging from 0.9 to 1.4 nm^2. Also, the total concentration of boron atoms required for initiating the formation of the monolayers is higher compared to the Ag(111) surface, due to the dissolution of the boron atoms with Au. The inspection of the STM images for the extended borophene phases reveals a periodicity of around 0.66 nm, with slight resemblance with the S1 phase as obtained on Ag(111) surface.

In order to understand the growth mechanism of boron layers on Au surface, DFT computations of the adsorption energetics of boron atoms on the surface were performed. The subsurface near the topmost layer along the step edges was found to offer the more favorable sites than both the surface locations and the bulk interior. In the presence of greater surface strain in Au, the subsurface diffusion of the boron atoms is energetically favored, allowing the formation a stronger Au-B interaction. The smaller boron clusters also favor the subsurface configurations by expelling the surface Au atoms. These clusters would eventually extend to the formation of the larger boron island with more boron atom deposition, removing the surface Au atoms. Figure 2.11 shows the computed energetics for the B diffusion on the surface and subsurface locations.

2.6 Interlayer Stacking of the Borophene Layers

The interlayer stacking possibilities and stabilities of borophene phases show large variations, depending upon the HD and the nonplanar arrangement of the boron atoms [56, 58, 64]. The monolayers could form interlayer bonds varying from trivial two-center covalent bonds to complex multicenter connections or prefer van der Waals interactions. Figures 2.12 and 2.13 shows the most preferred bilayer possibilities of different borophene phases, while in both sheets with zero HD, around half of the boron atoms per unit cell are involved in interlayer B-B bonds (bond density, BD = ½). The planar triangular phase (δ_6 sheet) gets rid of its one excess electron left behind in boron atoms at the hexagonal center upon the formation of the interlayer B-B bond [64, 65]. The bilayer sheet attains a similar

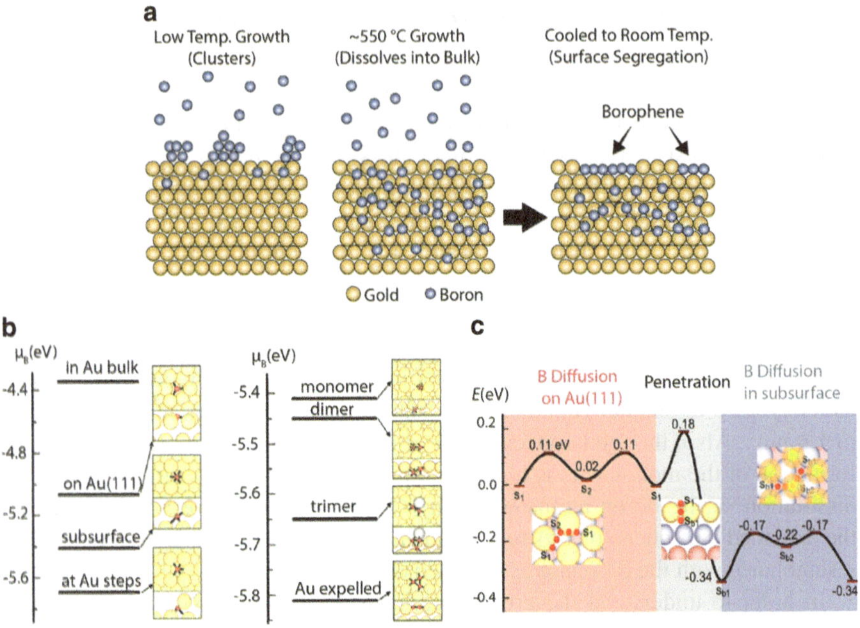

Fig. 2.11 (**a**) The growth mechanism of boron layers on Au surface at three different temperatures [10]. The (**b**) DFT calculations for the energy of boron atoms at different surface and subsurface locations on Au (111) surface and bulk. (**c**) The minimum energy path for the boron atom diffusion on the surface, penetration, and diffusion in subsurface. (Reprinted from ref. [10])

electron count as the AA-stacked graphene bilayer, such that intersection of the *p* bands oriented perpendicular to the hexagonal plane forms two Dirac cones at the Fermi region (Fig. 2.12a–b). With a slight increase in HD to near 1/16, the BD reduces to around ¼ [58]. At further increase of the HD such as in the α-boron bilayer (HD of around 1/9), interlayer B-B bonds are not observed, since the sheet is isoelectronic to graphene. Finally, the graphitic boron sheet (1/3 HD) prefer multicenter B-B bonds between the layers (Fig. 2.13) resembling the three-centered two-electron bonding patterns in several polyhedral boranes and three-dimensional allotropes to satisfy the skeletal electron count.

The relative stability of the bilayer sheets in comparison to the monolayer phases with and without the different metal templates is given in Table 2.2. Without the metal templates, the highest stability is obtained for α_8 sheet with the HD of 1/12, whereas the lowest binding energy is for δ_4 and δ_3 bilayers. On the metal templates, the interlayer interactions are affected by the overlap of the first layer with the metal surface. Most of the bilayers are destabilized compared to the corresponding monolayers on Cu surface. The Ag surface also shows a similar trend. The bilayers formed from the monolayers having higher interaction with the metal surface have lower stability, whereas the reduction in the interaction with the metal surface enhances the stability. However, for the Au surface, both the monolayers with higher

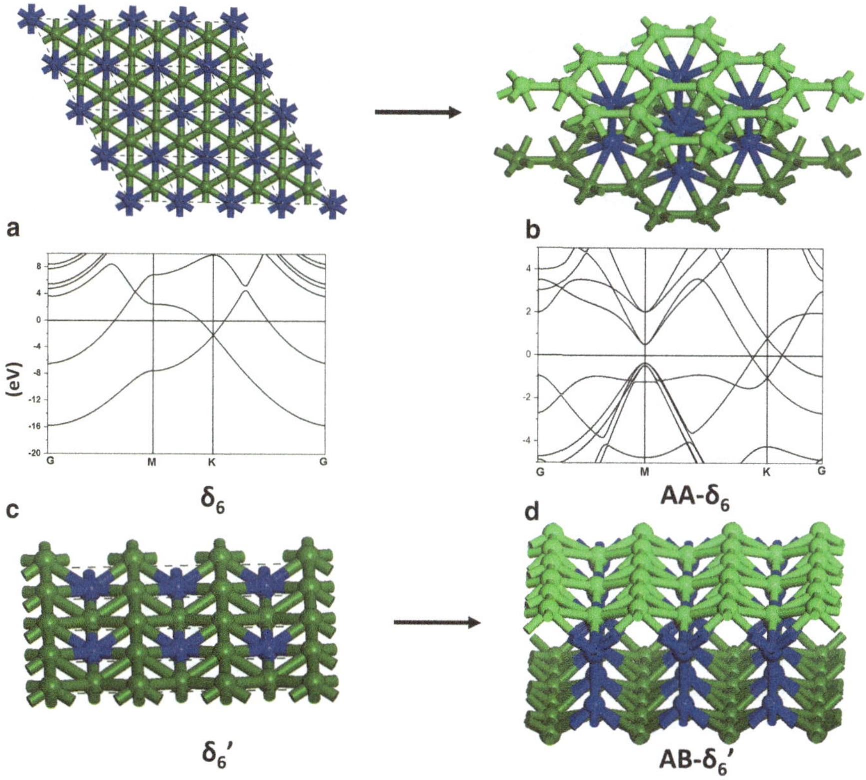

Fig. 2.12 Bilayer structures of boron sheets with zero HD. (**a**) and (**b**) Structure and band structure of monolayer δ_6 and AA-stacked δ_6 sheet. (**c**) δ_6' sheet. (**d**) AB-stacked δ_6' sheet

and lower HD get stabilized. The β_5 and β_{12} bilayers have almost similar stability as the corresponding monolayers, while for δ_6 and α_8 bilayer sheets, the stability is highest.

2.7 The Growth Mechanism of Borophene Phases

Having the thermodynamic data known for the stability of boron phases on different metal surfaces helps in understanding the growth mechanism of boron phases. The formation of 2D boron phases on the metal templates initiates with the formation of 2D boron clusters containing triangular units as the building blocks. Therefore, the nucleation of several 2D metal clusters starting from single boron atoms to around 20 atoms was studied on Cu(111) surface by Liu et al. [66]. The planar clusters containing only triangular B_3 units are found to be the most stable isomers with up to ten boron atoms. The formation of hexagonal hole initiated with the cluster

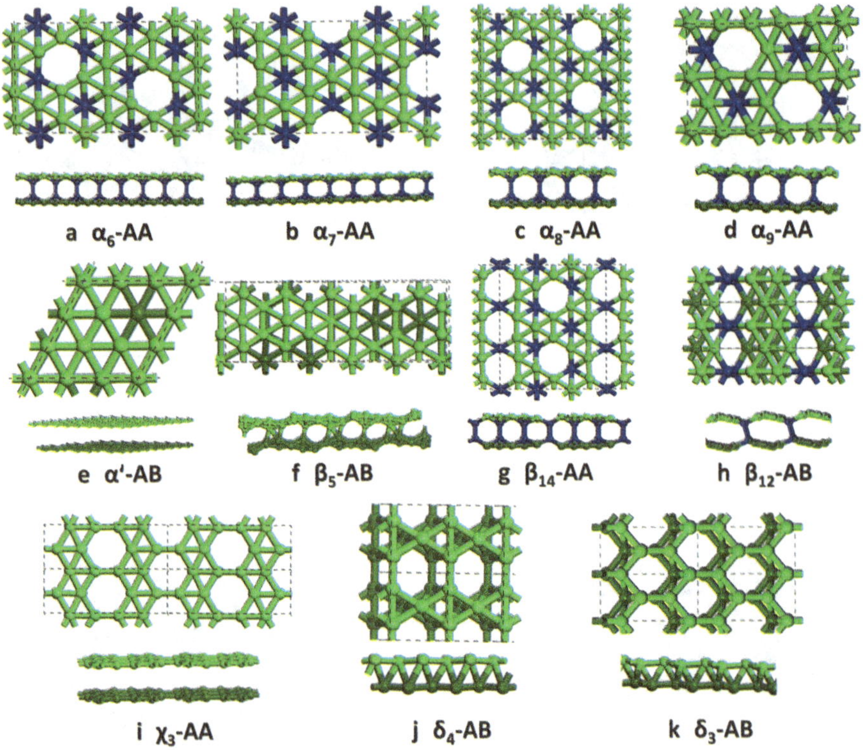

Fig. 2.13 The bilayer stacking possibilities of different monolayer borophenes with varying hole densities, HD. The interlayer bonds are denoted in blue. The bilayer arrangements of the layers are indicated by AA or AB stacking. (Reproduced from ref. [58])

having 11 boron atoms. However, with 12 and more boron atoms, the large number of isomers is obtained with almost similar stability. In most of these clusters, the belt-like elongated isomers containing triangular networks show higher stability. The formation of hexagonal holes is found to be favored, when it is enclosed by a sufficient number of the triangular units surrounding it. The corresponding three-dimensional isomers have lower stability. The 2D clusters undergo slight buckling when adsorbed on the surface, thereby effecting their electronic structures. The in-plane σ-orbitals undergo slight destabilization, whereas the π-orbitals gets lower in energy. The electron density moves from the in-plane molecular orbitals to the π-orbitals centered on the peripheral boron atoms. The energy per boron atom decreases initially with increase in N up to around 12 boron atoms. Beyond that, with the addition of more boron atoms, the formation of energy attains a saturation limit corresponding to the extended boron sheet. Once the metal template has sufficiently large islands of 2D boron clusters, the formation of 2D boron phases continues by joining of these smaller clusters. The molecular dynamics simulations of the boron clusters on the three metal surfaces (Cu, Ag, and Au) provide a similar

Table 2.2 The relative binding energies of boron sheets in meV/atom with respect to α-boron sheet

Sheet	HD	ΔBE_1 (meV/atom)	ΔBE_2 (meV/atom)			BD
			Cu	Ag	Au	
δ_6-AA	**0**	−29	–	69	44	2/6
δ_6'	0	96	50	13	21	
δ_6'-AB	**0**	−109	69	−42	−30	2/8
α_8	1/12 (0.083)	59	–	5	17	
α_8-AA	**1/12 (0.083)**	−146	–	−70	−48	24/88
α	1/9 (0.111)	0	0	0	0	–
β_5	2/15 (0.133)	1	−29	−28	6	–
β_5-AA	**2/15 (0.133)**	−27	63	−12	8	–
β_5-AB	**2/15 (0.133)**	−64	91	−40	15	Multicenter
β_{12}	1/6(0.167)	54	−78	−49	16	–
β_{12}-AA	**1/6(0.167)**	15	112	39	50	–
β_{12}-AB	**1/6(0.167)**	5	92	19	28	2/10

ΔBE_1 denotes the binding energy for the freestanding boron sheets, and ΔBE_2 is the binding energy on metal surfaces. BD denotes bond density, defined as a number of atoms involved in B-B bond formation/total number of atoms per unit cell. The HD of the bilayer borophene sheets are denoted in bold

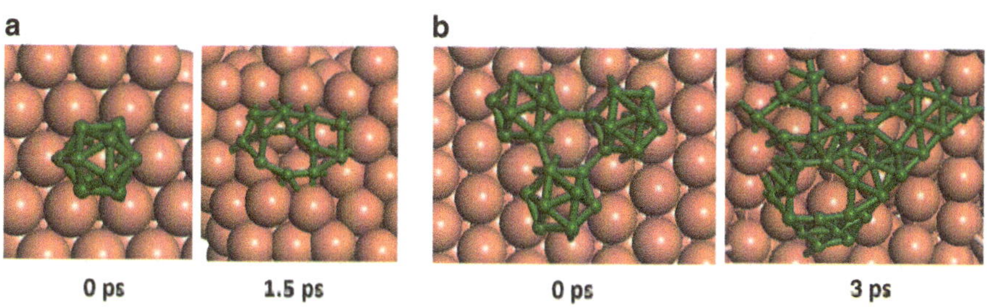

Fig. 2.14 The conformational changes of I_h-boron clusters of different sizes during simulations on Cu(111) surfaces at 1000 K temperature. (Reprinted from ref. [58]). The planar structures are found to be more stable monolayers with the metal surface

overview [58]. When the icosahedral clusters of different sizes are simulated on the metal templates, the rearrangement of the boron atoms to the planar forms is observed. Figure 2.14 shows the structural changes of the boron clusters on Cu(111) surface simulated at 1000 K temperature. For the continuing growth of an established borophene phase, the nanoreactor theory, originally developed for graphene [67], becomes well applicable as was recently demonstrated [68], in good accord with experiments. The formation of bilayers should initiate only after the complete formation of monolayers and depends upon the relative stability and interaction of the monolayers with the metal surface.

2.8 Conclusions and Future Outlook

One can sum up by saying that 2D boron development is a remarkably satisfying, even learning, example in materials science, where decades after semi-intuitive musings on such material possibility, the quantitative structure and synthesis predictions culminated in experimental discoveries. Its notable polymorphism [22, 69] and electronic structure, rich with topological features [55, 70, 71] and promise of superconductivity, motivate the accelerating research. Based upon the first-principles simulations, the most stable 2D boron polymorphs are studied with several future applications in mind. While the presence of hexagonal holes along with the triangular network in the borophene phases are found to enhance the in-plane elasticity and ductility [72], for interconnects in flexible/stretchable electronics, the fluxional multicenter bonding nature provides suitable platform for catalyzing a large number of surface reactions and metal ion mobility [73, 74]. Particularly, the β_{12} sheet with 1/6 HD is theoretically estimated to have a lower diffusion barrier for the Li and Na ions, thereby predicting excellent materials for devising Li and Na ion batteries [74]. Moreover, for several of these polymorphs, the analysis of their phonon spectra and electron-phonon coupling constants reveals superconducting properties with the critical temperature around 10–20 K [55]. Low-damped plasmons sustainable up to visible range [75], along with the unique optical reflectance of borophene layers [76], surpassing in theory even metallic silver, offer applications in reflecting and light-protective coatings.

Though the potential physics and even electronics and plasmonics applications are tantalizing, the challenges remain significant in simplifying the synthetic routes, mostly limited so far to sophisticated beam epitaxy. Separating boron monoatomic layers from the presently all-metal substrates, or finding the ways to grow borophene on insulators-semiconductors, would allow more direct and precise characterization of its intriguing properties, bringing the research to a new level.

References

1. B. Albert, H. Hillebrecht, Boron: Elementary challenge for experimenters and theoreticians. Angew. Chem., Int. Ed **48**, 8640–8668 (2009)
2. E.D. Jemmis, M.M. Balakrishnarajan, Polyhedral boranes and elemental boron: Direct structural relations and diverse electronic requirements. J. Am. Chem. Soc. **123**, 4324–4330 (2001)
3. T. Ogitsu, E. Schwegler, G. Galli, β-Rhombohedral boron: At the crossroads of the chemistry of boron and the physics of frustration. Chem. Rev. **113**, 3425–3449 (2013)
4. A.J. Mannix, X.-F. Zhou, B. Kiraly, J.D. Wood, D. Alducin, B.D. Myers, X. Liu, B.L. Fisher, U. Santiago, J.R. Guest, M.J. Yacaman, A. Ponce, A.R. Oganov, M.C. Hersam, N.P. Guisinger, Synthesis of borophenes: Anisotropic, two-dimensional boron polymorphs. Science **350**, 1513–1516 (2015)
5. B. Feng, J. Zhang, Q. Zhong, W. Li, S. Li, H. Li, P. Cheng, S. Meng, L. Chen, K. Wu, Experimental realization of two-dimensional boron sheets. Nat. Chem. **8**, 563–568 (2016)
6. W.-L. Li, X. Chen, T. Jian, T.-T. Chen, J. Li, L.-S. Wang, From planar boron clusters to borophenes and metalloborophenes. Nat. Rev. Chem. **1**, 0071 (2017)

7. A.P. Sergeeva, I.A. Popov, Z.A. Piazza, W.-L. Li, C. Romanescu, L.-S. Wang, A.I. Boldyrev, Understanding boron through size-selected clusters: Structure, chemical bonding, and fluxionality. Acc. Chem. Res. **47**, 1349–1358 (2014)
8. E.D. Jemmis, D.L. Prasad, Unknowns in the chemistry of boron. Curr. Sci. **95**, 1277–1283 (2008)
9. N. Karmodak, E.D. Jemmis, Fragment approach to the electronic structure of τ-boron allotrope. Phys. Rev. B **95**, 165128 (2017)
10. B. Kiraly, X. Liu, L. Wang, Z. Zhang, A.J. Mannix, B.L. Fisher, B.I. Yakobson, M.C. Hersam, N.P. Guisinger, Borophene synthesis on au(111). ACS Nano **13**, 3816–3822 (2019)
11. Z. Qing, Z. Jin, C. Peng, F. Baojie, L. Wenbin, S. Shaoxiang, L. Hui, M. Sheng, C. Lan, W. Kehui, Metastable phases of 2D boron sheets on Ag(1 1 1). J. Phys.: Condens. Matter **29**, 095002 (2017)
12. G. Tai, T. Hu, Y. Zhou, X. Wang, J. Kong, T. Zeng, Y. You, Q. Wang, Synthesis of atomically thin boron films on copper foils. Angew. Chem., Int. Ed. **54**, 15473–15477 (2015)
13. R. Wu, I.K. Drozdov, S. Eltinge, P. Zahl, S. Ismail-Beigi, I. Božović, A. Gozar, Large-area single-crystal sheets of borophene on cu(111) surfaces. Nat. Nanotechnol. **14**, 44–49 (2019)
14. I. Boustani, Systematic ab initio investigation of bare boron clusters: Determination of the geometry and electronic structures of B_n (n= 2–14). Phys. Rev. B **55**, 16426 (1997)
15. Y. Liu, E.S. Penev, B.I. Yakobson, Probing the synthesis of two-dimensional boron by first-principles computations. Angew. Chem. **125**, 3238–3241 (2013)
16. H. Tang, S. Ismail-Beigi, Novel precursors for boron nanotubes: The competition of two-center and three-center bonding in boron sheets. Phys. Rev. Lett. **99**, 115501 (2007)
17. X. Wu, J. Dai, Y. Zhao, Z. Zhuo, J. Yang, X.C. Zeng, Two-dimensional boron monolayer sheets. ACS Nano **6**, 7443–7453 (2012)
18. X. Yu, L. Li, X.-W. Xu, C.-C. Tang, Prediction of two-dimensional boron sheets by particle swarm optimization algorithm. J. Phys. Chem. C **116**, 20075–20079 (2012)
19. Z. Zhang, Y. Yang, G. Gao, B.I. Yakobson, Two-dimensional boron monolayers mediated by metal substrates. Angew. Chem., Int. Ed. **54**, 13022–13026 (2015)
20. Z. Zhang, S.N. Shirodkar, Y. Yang, B.I. Yakobson, Gate-voltage control of borophene structure formation. Angew. Chem., Int. Ed. **56**, 15421–15426 (2017)
21. S.N. Shirodkar, E.S. Penev, B.I. Yakobson, Honeycomb boron: Alchemy on aluminum pan? Sci. Bull. **63**, 270–271 (2018)
22. E.S. Penev, S. Bhowmick, A. Sadrzadeh, B.I. Yakobson, Polymorphism of two-dimensional boron. Nano Lett. **12**, 2441 2445 (2012)
23. E.D. Jemmis, E.G. Jayasree, Analogies between boron and carbon. Acc. Chem. Res. **36**, 816–824 (2003)
24. I. Boustani, New quasi-planar surfaces of bare boron. Surf. Sci. **370**, 355–363 (1997)
25. I. Boustani, New convex and spherical structures of bare boron clusters. J. Solid State Chem. **133**, 182–189 (1997)
26. H. Hogeveen, P.W. Kwant, J. Postma, P.T. van Duynen, Electronic spectra of pyramidal dications, $(CCH_3)_6{}^{2+}$ and $(CH)_6{}^{2+}$. Tetrahedron Lett. **15**, 4351–4354 (1974)
27. E.D. Jemmis, P.V.R. Schleyer, Aromaticity in three dimensions. 4. Influence of orbital compatibility on the geometry and stability of capped annulene rings with six interstitial electrons. J. Am. Chem. Soc **104**, 4781–4788 (1982)
28. M. Böyükata, C. Özdoğan, Z.B. Güvenç, Effects of hydrogen hosting on cage structures of boron clusters: Density functional study of B_mH_n (m = 5–10 and n $\leq$ m) complexes. Phys. Scr. **77**, 025602 (2008)
29. N. Gonzalez Szwacki, C.J. Tymczak, $B_{12}H_n$ and $B_{12}F_n$: Planar vs icosahedral structures. Nanoscale Res. Lett. **7**, 1–6 (2012)
30. N. Karmodak, R. Chaliha, E.D. Jemmis, Overlap of radial dangling orbitals controls the relative stabilities of polyhedral B_nH_{n-x} isomers (n = 5–12, x = 0 to n – 1). Inorg. Chem. **58**, 3627–3634 (2019)
31. J.K. Olson, A.I. Boldyrev, Planar to 3D transition in the B6Hy anions. J. Phys. Chem. A **117**, 1614–1620 (2013)

32. A.P. Sergeeva, Z.A. Piazza, C. Romanescu, W.-L. Li, A.I. Boldyrev, L.-S. Wang, B_{22}^- and B_{23}^-: All-boron analogues of anthracene and Phenanthrene. J. Am. Chem. Soc. **134**, 18065–18073 (2012)
33. N. Gonzalez Szwacki, A. Sadrzadeh, B.I. Yakobson, B_{80} fullerene: An Ab-initio prediction of geometry, stability, and electronic structure. Phys. Rev. Lett. **98**, 166804 (2007)
34. M.S. Dresselhaus, G. Dresselhaus, P.C. Eklund, *Science of Fullerenes and Carbon Nanotubes: Their Properties and Applications* (Academic press, 1996)
35. E.S. Penev, V.I. Artyukhov, F. Ding, B.I. Yakobson, Unfolding the fullerene: Nanotubes, graphene and poly-elemental varieties by simulations. Adv. Mater. **24**, 4956–4976 (2012)
36. D.L.V.K. Prasad, E.D. Jemmis, Stuffing improves the stability of Fullerenelike boron clusters. Phys. Rev. Lett. **100**, 165504 (2008)
37. D.L.V.K. Prasad, E.D. Jemmis, Stuffed fullerenelike boron carbide nanoclusters. Appl. Phys. Lett. **96**, 023108 (2010)
38. W. Huang, L.-S. Wang, Probing the 2D to 3D structural transition in gold cluster anions using argon tagging. Phys. Rev. Lett. **102**, 153401 (2009)
39. Y. Ohishi, K. Kimura, M. Yamaguchi, N. Uchida, T. Kanayama, Formation of hydrogenated boron clusters in an external quadrupole static attraction ion trap. J. Chem. Phys. **128**, 124304 (2008)
40. Y. Ohishi, K. Kimura, M. Yamaguchi, N. Uchida, T. Kanayama, Synthesis and formation mechanism of hydrogenated boron clusters $B_{12}H_n$ with controlled hydrogen content. J. Chem. Phys. **133**, 074305 (2010)
41. D.Y. Zubarev, A.I. Boldyrev, Comprehensive analysis of chemical bonding in boron clusters. J. Comput. Chem. **28**, 251–268 (2007)
42. D.Y. Zubarev, A.I. Boldyrev, Developing paradigms of chemical bonding: Adaptive natural density partitioning. Phys. Chem. Chem. Phys. **10**, 5207–5217 (2008)
43. B. Kiran, S. Bulusu, H.-J. Zhai, S. Yoo, X.C. Zeng, L.-S. Wang, Planar-to-tubular structural transition in boron clusters: B_{20} as the embryo of single-walled boron nanotubes. Proc. Natl. Acad. Sci. U. S. A. **102**, 961–964 (2005)
44. N. Karmodak, E.D. Jemmis, Exohedral complexation of B_{40}, C_{60} and Arenes with transition metals: A comparative DFT study. Chem. Asian J. **11**, 3350–3354 (2016)
45. H.-J. Zhai, Y.-F. Zhao, W.-L. Li, Q. Chen, H. Bai, H.-S. Hu, Z.A. Piazza, W.-J. Tian, H.-G. Lu, Y.-B. Wu, Y.-W. Mu, G.-F. Wei, Z.-P. Liu, J. Li, S.-D. Li, L.-S. Wang, Observation of an all-boron fullerene. Nat. Chem. **6**, 727–731 (2014)
46. Z.A. Piazza, H.S. Hu, W.L. Li, Y.F. Zhao, J. Li, L.S. Wang, Planar hexagonal B_{36} as a potential basis for extended single-atom layer boron sheets. Nat. Commun. **5**, 3113 (2014)
47. S. Suehara, T. Aizawa, T. Sasaki, Graphenelike surface boron layer: Structural phases on transition-metal diborides (0001). Phys. Rev. B **81**, 085423 (2010)
48. M.E. Jones, R.E. Marsh, The preparation and structure of magnesium boride, MgB_2. J. Am. Chem. Soc. **76**, 1434–1436 (1954)
49. D.L.V.K. Prasad, E.D. Jemmis, Boron and MgB_2 analogs of fullerenes and carbon nanotubes: A density functional theory study. J. Mol. Struc.: THEOCHEM **771**, 111–115 (2006)
50. H. Tang, S. Ismail-Beigi, Self-doping in boron sheets from first principles: A route to structural design of metal boride nanostructures. Phys. Rev. B **80**, 134113 (2009)
51. N. Karmodak, E.D. Jemmis, The role of holes in borophenes: An Ab initio study of their structure and stability with and without metal templates. Angew. Chem., Int. Ed **56**, 10093–10097 (2017)
52. J. Kunstmann, A. Quandt, Broad boron sheets and boron nanotubes: An ab initio study of structural, electronic, and mechanical properties. Phys. Rev. B **74**, 035413 (2006)
53. H. Tang, S. Ismail-Beigi, First-principles study of boron sheets and nanotubes. Phys. Rev. B **82**, 115412 (2010)
54. H. Shu, F. Li, P. Liang, X. Chen, Unveiling the atomic structure and electronic properties of atomically thin boron sheets on an ag(111) surface. Nanoscale **8**, 16284–16291 (2016)
55. E.S. Penev, A. Kutana, B.I. Yakobson, Can two-dimensional boron superconduct? Nano Lett. **16**, 2522–2526 (2016)

56. N. Gao, X. Wu, X. Jiang, Y. Bai, J. Zhao, Structure and stability of bilayer borophene: The roles of hexagonal holes and interlayer bonding. FlatChem **7**, 48–54 (2018)
57. L.Z. Zhang, Q.B. Yan, S.X. Du, G. Su, H.J. Gao, Boron sheet adsorbed on metal surfaces: Structures and electronic properties. J. Phys. Chem. C **116**, 18202–18206 (2012)
58. N. Karmodak, E.D. Jemmis, Metal templates and boron sources controlling Borophene structures: An ab initio study. J. Phys. Chem. C **122**, 2268–2274 (2018)
59. Z. Zhang, A.J. Mannix, Z. Hu, B. Kiraly, N.P. Guisinger, M.C. Hersam, B.I. Yakobson, Substrate-induced nanoscale undulations of Borophene on silver. Nano Lett. **16**, 6622–6627 (2016)
60. X. Liu, Z. Zhang, L. Wang, B.I. Yakobson, M.C. Hersam, Intermixing and periodic self-assembly of borophene line defects. Nat. Mater. **17**, 783–788 (2018)
61. W. Li, L. Kong, C. Chen, J. Gou, S. Sheng, W. Zhang, H. Li, L. Chen, P. Cheng, K. Wu, Experimental realization of honeycomb borophene. Sci. Bull. **63**, 282–286 (2018)
62. X. Liu, L. Wang, S. Li, M.S. Rahn, B.I. Yakobson, M.C. Hersam, Geometric imaging of borophene polymorphs with functionalized probes. Nat. Commun. **10**, 1642 (2019)
63. X. Sun, X. Liu, J. Yin, J. Yu, Y. Li, Y. Hang, X. Zhou, M. Yu, J. Li, G. Tai, W. Guo, Two-dimensional boron crystals: Structural stability, tunable properties, fabrications and applications. Adv. Funct. Mater. **27**, 1603300 (2017)
64. F. Ma, Y. Jiao, G. Gao, Y. Gu, A. Bilic, Z. Chen, A. Du, Graphene-like two-dimensional ionic boron with double Dirac cones at ambient condition. Nano Lett. **16**, 3022–3028 (2016)
65. X.-F. Zhou, X. Dong, A.R. Oganov, Q. Zhu, Y. Tian, H.-T. Wang, Semimetallic two-dimensional boron allotrope with massless Dirac fermions. Phys. Rev. Lett. **112**, 085502 (2014)
66. H. Liu, J. Gao, J. Zhao, From boron cluster to two-dimensional boron sheet on cu(111) surface: Growth mechanism and hole formation. Sci. Rep. **3**, 3238 (2013)
67. V.I. Artyukhov, Y. Liu, B.I. Yakobson, Equilibrium at the edge and atomistic mechanisms of graphene growth. Proc. Natl. Acad. Sci. **109**, 15136–15140 (2012)
68. Z. Zhang, A.J. Mannix, X. Liu, Z. Hu, N.P. Guisinger, M.C. Hersam, B.I. Yakobson, Near-equilibrium growth from borophene edges on silver. Sci. Adv **5**, eaax0246 (2019)
69. Z. Zhang, E.S. Penev, B.I. Yakobson, Two-dimensional materials: Polyphony in B flat. Nat. Chem. **8**, 525–527 (2016)
70. B. Feng, O. Sugino, R.-Y. Liu, J. Zhang, R. Yukawa, M. Kawamura, T. Iimori, H. Kim, Y. Hasegawa, H. Li, L. Chen, K. Wu, H. Kumigashira, F. Komori, T.-C. Chiang, S. Meng, I. Matsuda, Dirac fermions in Borophene. Phys. Rev. Lett. **118**, 096401 (2017)
71. H. Zhang, Y. Xie, Z. Zhang, C. Zhong, Y. Li, Z. Chen, Y. Chen, Dirac nodal lines and tilted semi-Dirac cones coexisting in a striped boron sheet. J. Phys. Chem. Lett. **8**, 1707–1713 (2017)
72. Z. Zhang, Y. Yang, E.S. Penev, B.I. Yakobson, Elasticity, flexibility, and ideal strength of Borophenes. Adv. Funct. Mater. **27**, 1605059 (2017)
73. C. Liu, Z. Dai, J. Zhang, Y. Jin, D. Li, C. Sun, Two-dimensional boron sheets as metal-free catalysts for hydrogen evolution reaction. J. Phys. Chem. C **122**, 19051–19055 (2018)
74. X. Zhang, J. Hu, Y. Cheng, H.Y. Yang, Y. Yao, S.A. Yang, Borophene as an extremely high capacity electrode material for Li-ion and Na-ion batteries. Nanoscale **8**, 15340–15347 (2016)
75. Y. Huang, S.N. Shirodkar, B.I. Yakobson, Two-dimensional boron polymorphs for visible range Plasmonics: A first-principles exploration. J. Am. Chem. Soc. **139**, 17181–17185 (2017)
76. S. Gupta, S.N. Shirodkar, A. Kutana, B.I. Yakobson, In pursuit of 2D materials for maximum optical response. ACS Nano **12**, 10880–10889 (2018)

Chapter 3
Synthesis of Borophene

Baojie Feng, Lan Chen, and Kehui Wu

3.1 Introduction

As the fifth element in the periodic table, boron is located between metallic beryllium and nonmetallic carbon. The ground-state configuration of boron is $[He]2s^22p^1$, which means that boron has three valence electrons and four atomic orbitals. It is obvious that the number of atomic orbitals is larger than the number of available valence electrons, which prevents fulfillment of the octet rule. In other words, there are too few valence electrons to occupy all the orbitals of the classical two-center two-electron chemical bonds, which causes the electron-deficient state in boron. As a result, various types of chemical bonds can coexist in boron, including conventional two-center two-electron bonds and unconventional delocalized multicenter two-electron bonds. The complexity of the chemical bonds in boron gives rise to tens of bulk polymorphs that consist of interconnected B_{12} icosahedral cages [1, 2]. The situation is similar in borophene, in which the thickness of boron is reduced to a single atomic layer. Theoretically, it is commonly accepted that freestanding borophene consists of a triangular lattice with hollow hexagons (HHs) to balance out the two-center and multicenter bonds [3], and the stability of borophene is directly related to the HH density (ν). The different arrangements of the HHs in borophene produce a number of polymorphs, and, surprisingly, most of these polymorphs are metallic, in contrast to the semiconducting or insulating nature of bulk boron [3, 4].

Unfortunately, despite the polymorphism of boron, none of the polymorphs have a graphite-like layered structure. As a result, it is impossible to obtain borophene using conventional mechanical exfoliation techniques. Theoretically, the energy per

B. Feng · L. Chen · K. Wu (✉)
Institute of Physics, Chinese Academy of Sciences, Beijing, China
e-mail: khwu@iphy.ac.cn

I. Matsuda, K. Wu (eds.), *2D Boron: Boraphene, Borophene, Boronene*,
https://doi.org/10.1007/978-3-030-49999-0_3

boron atom in borophene is much higher than that in bulk boron, which indicates that borophene is thermodynamically unstable. In other words, large numbers of boron atoms would preferentially form 3D bulk structures rather than the two-dimensional (2D) borophene. Furthermore, different borophene polymorphs with similar degrees of stability tend to coexist because of the intrinsic polymorphism of borophene; this will also make it difficult to synthesize large-scale single-phase borophene. As a result, the experimental synthesis of borophene has become a challenging task for researchers.

One powerful way to synthesize 2D materials is molecular beam epitaxy (MBE), which requires an ultrahigh vacuum. In MBE, the deposition and diffusion rates of the atoms on the surface can be controlled well via the temperatures of the sources and the substrates, respectively. When this control is combined with the catalysis effects of the substrates, it is possible to suppress the formation of 3D clusters, even if the 3D clusters are energetically more favorable. The powerful nature of MBE has enabled the realization of synthetic 2D materials that lack 3D parent structures, such as silicene and germanene [5, 6]. As a result, MBE is a good choice for synthesis of borophene. For growth of boron, one of the most important factors is the selection of the supporting substrate, which is crucial for the stabilization of borophene. The successful growth of graphene and silicene on metal substrates, such as Cu(111), Ag(111), and Pt(111), suggests that metal surfaces can assist in the growth of 2D materials, and these substrates may be applicable in the case of borophene. The feasibility of the use of metal substrates has already been studied theoretically by several groups before the experimental realization of borophene [7–9]. For example, Liu et al. reported that boron preferentially forms 2D flat structures on Cu(111) and Ag(111) surfaces and noted that a high nucleation barrier could impede the formation of 3D clusters [7]. Zhang et al. reported that borophene can be grown on a variety of metal substrates, including Mg, Al, Ti, Au, and Ag [8].

These pioneering theoretical works have stimulated major experimental efforts aimed at the synthesis of borophene. As expected, various borophene polymorphs have been synthesized successfully on metal substrates. In the following, we will briefly introduce and describe these experimental works based on the substrates used.

3.2 Ag(111)

The synthesis of borophene was first realized on Ag(111) substrates by two independent groups. By evaporating pure boron directly onto Ag(111) at 550 °C, Mannix et al. obtained atomically thin boron sheets, i.e., borophene [10]. Using scanning tunneling microscopy (STM), they observed two borophene phases that were dependent on the deposition rate and the substrate temperature: a homogeneous phase and a striped phase, as shown in Fig. 3.1(a–c). Higher growth temperatures favored the growth of the striped phase, thus suggesting that the striped phase is thermodynamically more stable. High-resolution STM measurements showed that

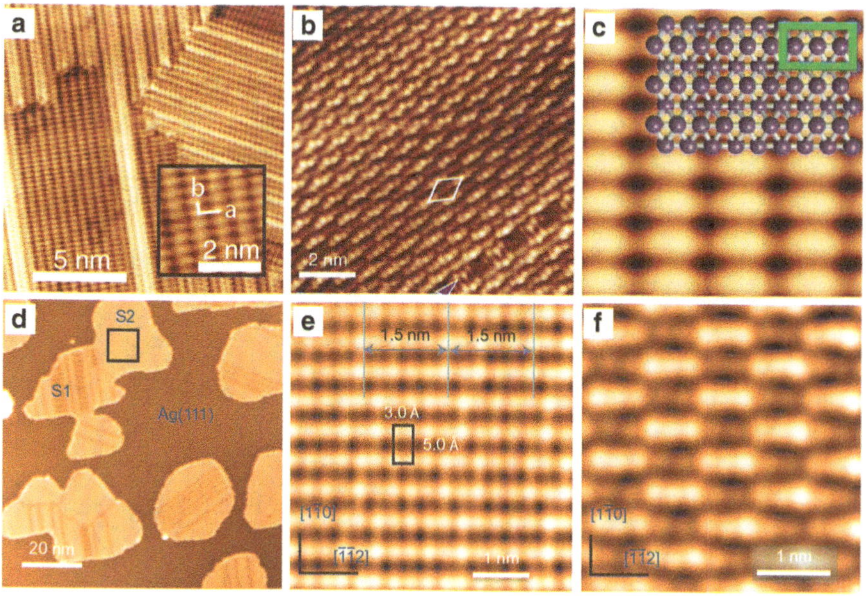

Fig. 3.1 STM images of different phases of borophene on Ag(111) surfaces. (**a–c**) (Data from Ref. [10]. Reprinted by permission from Mannix et al. 2015. Copyright (2015) American Association for the Advancement of Science. (**d–f**) Data from Ref. [11])

the homogeneous phase consists of periodic atomic chains (with 0.3 nm periodicity) with periodic vertical buckling, a short rhombohedral moiré pattern, and a long-range 1D moiré pattern. The striped phase has a rectangular lattice with lattice constants of 0.51 nm and 0.29 nm. Combining their results with theoretical calculations, they proposed that both phases have a close-packed borophene structure without HHs. These borophene sheets have intrinsic corrugations on the Ag(111) surface because of the instability of the freestanding form, thus resulting in the two phases shown in their STM images [10].

Almost simultaneously, Feng et al. reported similar experimental results for the synthesis of borophene, but their structural models were different [11]. They also obtained two phases of borophene, named S1 and S2 in their paper, as shown in Fig. 3.1(d and e). The S2 phase shown in their STM images closely resembles the homogeneous phase reported in the work of Mannix et al. However, despite the similar rectangular lattice shown in the STM images, the S1 phase has an additional 1D moiré pattern with periodicity of 1.5 nm (Fig. 3.1e); this 1D moiré pattern is absent from the striped phase in the work of Mannix et al. By annealing at a higher temperature, the S1 phase can be transformed into the S2 phase, which demonstrated that the S2 phase is more stable. The higher stability of this S2 phase has also been reported in subsequent experimental works [12–14]. By considering their results in combination with first-principles calculations, Feng et al. proposed that the two phases are triangular boron structures with HHs that were already

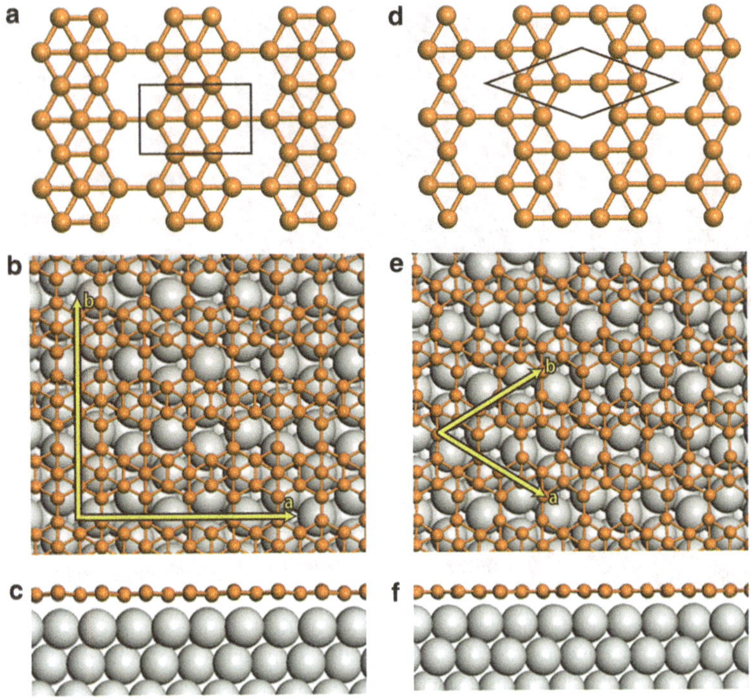

Fig. 3.2 Structural models of borophenes on Ag(111). (**a**) Freestanding β_{12} borophene. (**b**) and (**c**) Top and side views of β_{12} borophene on Ag(111), respectively. (**d–f**) The same images as shown in (**a–c**) for χ_3 borophene on Ag(111). Orange and gray balls represent the boron and silver atoms, respectively. (Reproduced from Ref. [11])

proposed in previous works [15]: S1 corresponds to the β_{12} sheet ($\nu = 1/6$) and S2 corresponds to the χ_3 sheet ($\nu = 1/5$), as illustrated in Fig. 3.2. Interestingly, both boron sheets remained planar without any obvious vertical corrugations, thus indicating a relatively weak interaction with the substrate. The calculated formation energy of the β_{12} sheet is slightly lower than that of the χ_3 sheet (Table 3.1), which agrees well with the greater stability of the S2 phase. In addition, they calibrated the atomic flux of boron based on the well-known Si(111)-B-$\sqrt{3} \times \sqrt{3}$ superstructure and confirmed that the densities of the boron atoms in the two borophene phases were approximately 1/6 and 1/5, which match the structural models of the β_{12} and χ_3 sheets, respectively.

The different structural models of borophene proposed by Mannix et al. and by Feng et al. leave an open question with regard to the correct structure of borophene on Ag(111). In the freestanding form, the β_{12} and χ_3 sheets are more stable than the triangular structure because of the existence of the HHs, which indicates that the β_{12} and χ_3 sheets have higher stability on Ag(111). Theoretical calculations with respect to the substrates were performed by Zhang et al. at almost the same time as the two experimental works, and they found that the β_{12} sheet is the ground state

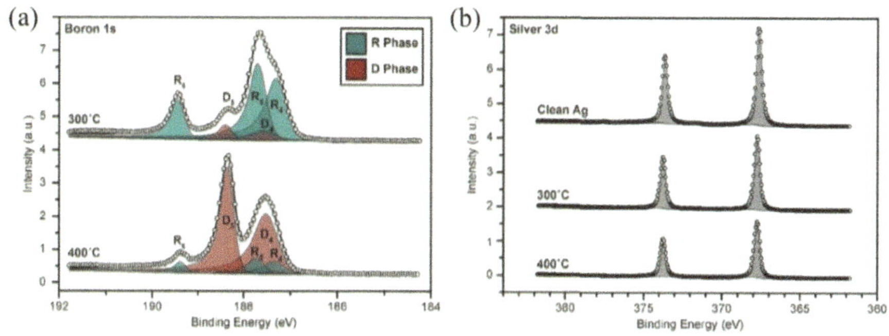

Fig. 3.3 High-resolution XPS results. (**a**) Boron 1 s core-level spectra for borophene samples grown at 300 and 400 °C. (**b**) Silver 3d core-level spectra of clean Ag(111) and borophene samples grown at 300 and 400 °C. Both plots display the raw data (dots), the fitted peak components (filled shapes), and fitted envelopes (gray lines). (Reproduced from [12]. Reprinted by permission from Campbell et al. 2018. Copyright (2018) American Chemical Society)

of borophene on Ag(111) via a global minimum search [16]. Subsequent theoretical calculations further supported the higher stability of the β_{12} and χ_3 sheets when compared with the triangular sheet on Ag(111) [17, 18].

The discussions above indicate that the β_{12} sheet is the correct structural model for the S1 phase reported by Feng et al. [11]. As noted above, the S1 phase has a 1D moiré pattern with a periodicity of 1.5 nm that is absent from the striped phase reported by Mannix et al. [10]. A closer inspection of the STM images of the two phases shows that the relative angle between the rectangular unit cell and the Ag(111) surface is different in the two cases. In the triangular structure case [10], the $[1\bar{1}0]$ direction of Ag(111) coincides with the short edge of the rectangle, while there is a 30° rotation in the case of the S1 phase [11]. Therefore, it is simple to assign the striped phase to the β_{12} sheet with a 30° rotation angle with respect to the S1 phase. This striped phase was also observed recently by Zhong et al. and has been labeled S3 [19].

X-ray photoelectron spectroscopy (XPS) is a powerful technique for study of the chemical bonding properties of borophene [12] and can provide useful information to support the structural models of borophene on Ag(111). The XPS spectra of the S1 and S2 phases are shown in Fig. 3.3. These spectra show three peaks for the S1 phase and two peaks for the S2 phase. The fitted intensity ratios show that the 300 °C sample is composed of ~90% S1 phase and ~10% S2 phase, whereas the 400 °C sample is composed of 10% S1 phase and 90% S2 phase. Furthermore, the ratios of the peak intensities are ~2:2:1 for the S1 phase and ~1:1 for the S2 phase, which are consistent with the predicted ratios of fourfold, fivefold, and sixfold coordinated boron atoms contained in the β_{12} and χ_3 borophene crystal structures. In contrast, the XPS spectra of the Ag 3d peaks are indistinguishable from those of clean Ag(111), which indicates the weak interaction between borophene and the Ag substrate.

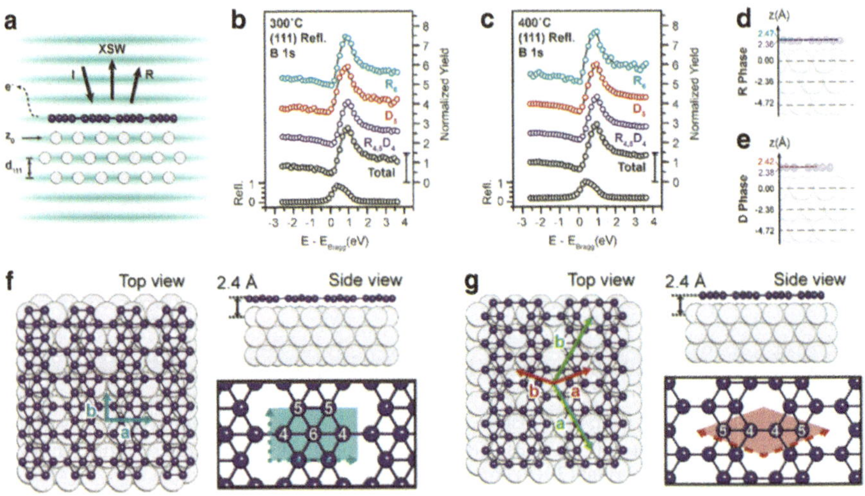

Fig. 3.4 (**a**) Schematic of XSW (blue lines) generated by interference between incident (I) and reflected (R) X-ray beams during Bragg diffraction from a Ag(111) single crystal with a χ_3 phase borophene layer. (**b**), (**c**) Data (circles) and model fittings (black lines) for the incident X-ray energy dependence of the X-ray reflectivity and the normalized B 1s photoelectron yields of samples grown at 300 and 400 °C, respectively. The total B 1s is shown along with the chemically distinct boron species from Fig. 3.3(**a**). (**d**), (**e**) Structural characterizations for the β_{12} phase and the χ_3 phase from XSW as derived from the 300 and 400 °C sample growths, respectively. (**f**), (**g**) Atomic models of Ag(111)-supported borophene in the β_{12} and χ_3 structures, respectively. The enlarged models show the low-energy electron diffraction (LEED)-derived β_{12} and χ_3 2D unit cells over each phase with the boron atomic coordination numbers labeled. (Reproduced from Ref. [12]. Reprinted by permission from Campbell et al. 2018. Copyright (2018) American Chemical Society)

These XPS spectra only provide evidence of the chemical coordination of the boron atoms in borophene, which represents an indirect approach to determination of the atomic structure of borophene. To determine the structures of borophene precisely, Campbell et al. used the X-ray standing wave (XSW) technique, which can enhance XPS with high spatial resolution, as illustrated in Fig. 3.4. The heights of the atoms within the distinct chemical environment can be determined by fitting of the peaks. Full details of the measurements and the fitting process used can be found in Ref. [12]. An analysis of the Ag $3d_{5/2}$ XSW-XPS results shows that all silver atoms located within the 10 Å sampling depth are unaffected by the presence of boron. For borophene, despite the distinct in-plane registry between the surface silver atoms and the boron atoms in both the β_{12} and χ_3 structures, the fitting results show that all the boron atoms are highly coplanar. The distance between the borophene and the topmost Ag(111) layer is 2.4 Å, with a negligible corrugation of $\Delta z \approx 0.1$ Å. These results indicate weak B-Ag hybridization, which is consistent with the recent discovery of metallicity and Dirac fermions in these structures [20–22].

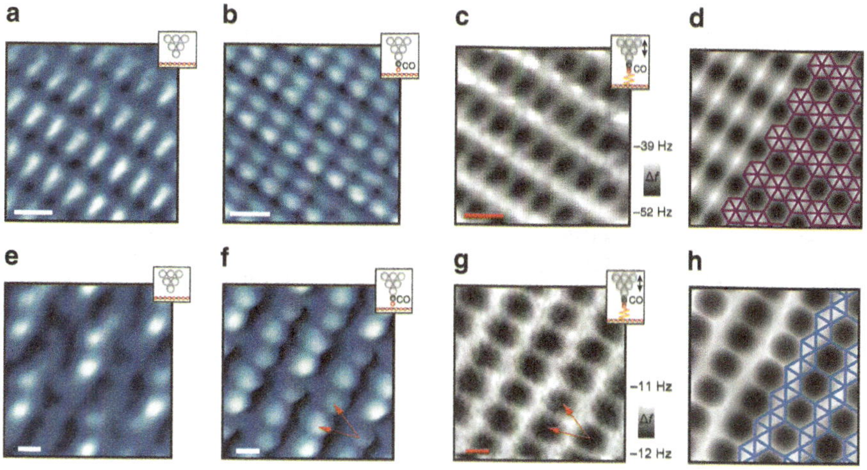

Fig. 3.5 (**a**, **b**) STM images of β_{12} borophene acquired with a bare tip and a CO-functionalized tip, respectively. (**c**) AFM image of β_{12} borophene acquired with a CO-functionalized tip. (**d**) Simulated CO-AFM image of β_{12} borophene. (**e**, **f**) STM images of χ_3 borophene acquired with a bare tip and a CO-functionalized tip, respectively. (**g**) AFM image of χ_3 borophene acquired with a CO-functionalized tip. (**h**) Simulated CO-AFM image of χ_3 borophene. (Reproduced from Ref. [13]. (By courtesy of Boris I. Yakobson))

Atomic force microscopy (AFM) with a CO-functionalized tip provides a powerful technique to resolve the atomic structures of materials and can provide higher spatial resolution than conventional STM. Recently, Liu et al. studied the atomic structure of borophene using a CO-functionalized AFM tip [13]. Figure 3.5(a, e) shows the STM images of the β_{12} and χ_3 borophenes, respectively, where only the periodicity of the two phases can be resolved. When the tip is adsorbed with a CO molecule, the STM spatial resolution is slightly improved, as illustrated in Fig. 3.5(b, f). The AFM mode can further improve the spatial resolution and directly resolve the HHs (Fig. 3.5(c, g)), which is consistent with simulated results (Fig. 3.5d, h, respectively). These results thus confirmed directly that the two observed borophene phases were the β_{12} and χ_3 sheets.

In addition to the dominant β_{12} and χ_3 borophenes on Ag(111), several metastable borophene phases are present on Ag(111). Zhong et al. reported two metastable phases, designated the S3 and S4 phases, and STM images of these phases are shown in Fig. 3.6. In combination with first-principles calculations, the S3 phase was identified as the β_{12} phase with a 30° rotation with respect to S1, while the S4 phase was the well-known α sheet, which was predicted to be the most stable freestanding borophene. As discussed above, the S3 phase is likely to be the striped phase that was reported by Mannix et al. [10]. These S3 and S4 phases always coexist with the S1 and S2 phases, but their ratios on Ag(111) are much smaller, which indicates the lower stability of these two phases. In fact, these two phases are hardly seen by low-energy electron diffraction (LEED) measurements because of their negligible ratios with respect to the S1 and S2 phases. According

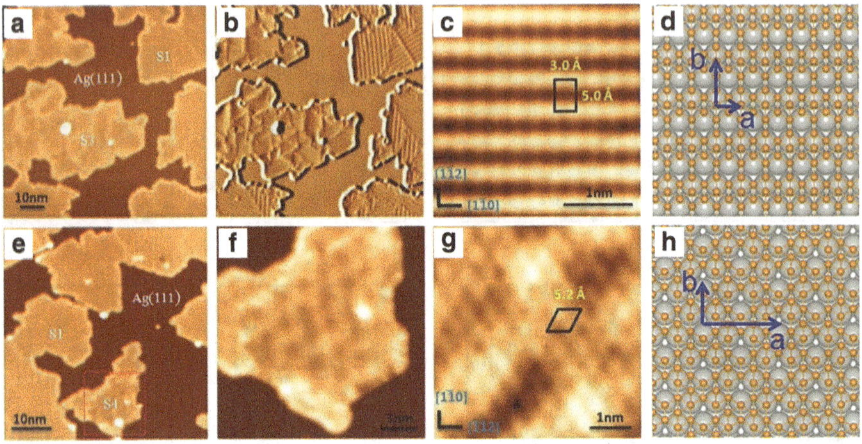

Fig. 3.6 STM images of two metastable 2D boron sheets on Ag(111). (**a**) STM topographic image of boron structures on Ag(111). The boron islands are labeled as the "S1" and "S3" phases. (**b**) Derivative STM image of (**a**). (**c**) High-resolution STM image of the S3 phase. The S3 unit cell is marked using a black rectangle. (**d**) Top view of the S3 model, which corresponds to a β_{12} sheet of 2D boron on the Ag(111) surface. (**e**) STM topographic image of boron structures on Ag(111). The boron islands are labeled as the "S1" and "S4" phases. Most boron islands shown in the image are of the S1 phase. (**f**) STM image acquired on the area marked by the red-dotted rectangle shown in (**d**). (**g**) High-resolution STM image of the S4 phase. The S4 unit cell is marked using a black rhombus. (**g**) Top view of the S4 model, which corresponds to the α sheet of 2D boron on Ag(111). (Reproduced from Ref. [19])

to the calculations, both S3 and S4 remain relatively flat on the Ag(111) surface, thus indicating their weak interaction with the substrate. The coexistence of these multiple borophene phases on Ag(111) confirms the polymorphism of borophene.

To provide a better understanding of the stability of borophene on Ag(111), it is desirable to compare the formation energies of the different phases. The calculation results for the formation energies are presented in Table 3.1. Among the freestanding borophenes, the most stable phase is the well-known α sheet with $\nu = 1/9$ [24]. However, on Ag(111), the total formation energy includes the adhesion energy of the substrates, and the total formation energy of the α sheet therefore becomes lower than that of the β_{12} and χ_3 sheets. As a result, on Ag(111), the β_{12} and χ_3 sheets become more stable than the α sheet, which is again in agreement with the experimental results. These results also demonstrate the power of MBE, which can be used to synthesize metastable and even unstable 2D materials in their freestanding forms.

In addition to the S1–S4 phases, Liu et al. [13] found several other borophene phases on Ag(111): β_{12}–22°, χ_3–6°, χ_3–9°, and χ_3–22°. Interestingly, the overall trend is that the borophene growth process transitions from rotationally commensurate phases to incommensurate phases at higher growth temperatures. Note that all phases of borophene on Ag(111) are β_{12} and χ_3 phases, except for a metastable α sheet. The difference that separates each of these borophenes is their rotation

Table 3.1 Formation energies for freestanding and epitaxial S1–S4 phases

	S1	S2	S3	S4
E_{FB}	6.23	6.19	6.23	6.27
E_{EB}	6.32	6.35	6.24	6.29

Extracted from Ref. [11, 19]
E_{FB} is the formation energy per atom of freestanding borophene. E_{EB} is the formation energy per atom for the epitaxial borophene on Ag(111)

angle with respect to the Ag(111) surface. The existence of these rotationally incommensurate phases provides corroborating evidence that the borophene layers are chemically discrete from the underlying Ag surface.

From the discussions above, we see that β_{12} and χ_3 borophenes are the predominant phases on Ag(111). Over a large growth temperature range, these two phases coexist on a Ag(111) surface with rich line defects, as shown in Fig. 3.7(a, b) [23]. These line defects arise from the β_{12}-β_{12}, β_{12}-χ_3, or χ_3-χ_3 domain boundaries. The registry of β_{12} and χ_3 sheets along the HH rows allows these sheets to connect seamlessly and thus exhibit an atomically smooth phase boundary (Fig. 3.7d). In other words, the line defects are formed by intermixing of the β_{12} and χ_3 rows.

The perfect lattice matching of the β_{12} and χ_3 sheets along the HH rows at the phase boundaries results in a negligible interface energy, in contrast to the more substantial interface energies in other configurations that are caused by large lattice mismatches (Fig. 3.7e). Furthermore, the two borophenes are highly anisotropic, which will lead to a well-defined defect formation. These defect-induced phases blur the distinction between the borophene crystals and the defects. Because the crystals are defined by their atomic ordering and structural periodicity, borophene domains containing periodic assemblies of β_{12} and χ_3 rows can be viewed equivalently as new borophene phases. For example, Fig. 3.7(e) shows a borophene domain that displays different regions that are defined by line defects with distinct periodic lengths and are separated by the white-dashed lines. Detailed analysis shows that two more borophene phases are formed at the domain boundaries with HH densities of $\nu = 4/21$ and 7/3, respectively. In addition, low-temperature measurements showed subtle electronic features that are consistent with a charge density wave and are modulated by the line defects. Recently, by depositing perylenetetracarboxylic dianhydride (PTCDA) molecules on borophene/Ag(111), Liu et al. produced lateral heterostructures with electronically abrupt interfaces [25]; their work may inform emerging efforts to integrate borophene into materials for nanoelectronic applications.

In addition to the large area borophene sheets on Ag(111), Zhang et al. found that reduction of the growth temperature and the chemical potential of boron can enhance the kinetic anisotropy in a manner that promotes the growth of ultra-narrow boron nanoribbons [26]. The formation of these boron nanoribbons originates from

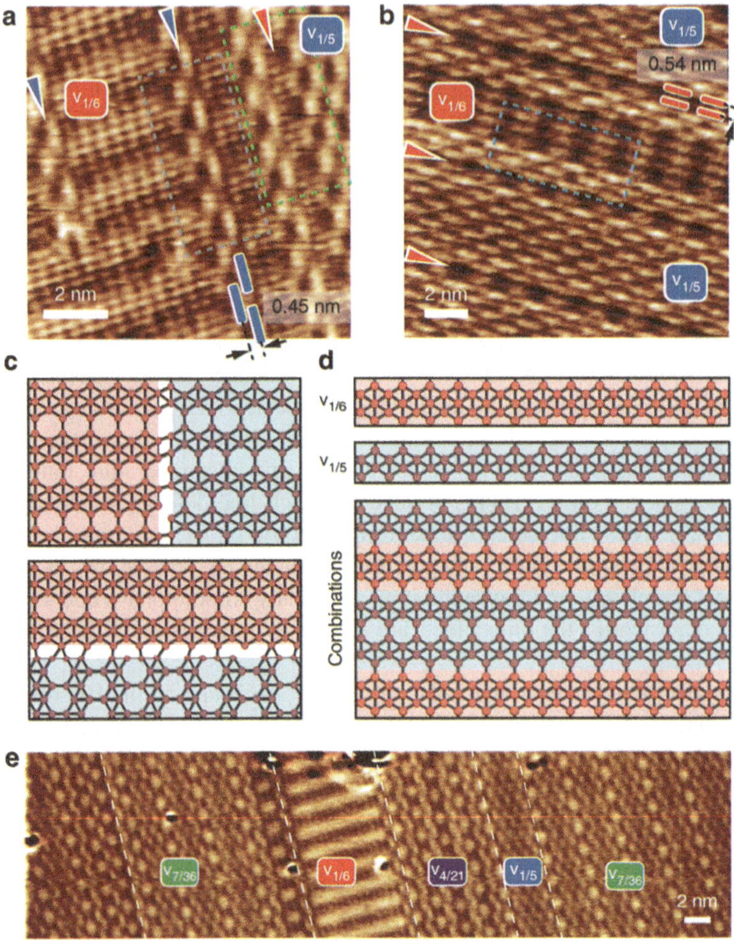

Fig. 3.7 (**a** and **b**) STM images of line defects in β_{12} and χ_3 borophene sheets, as indicated by the blue and red arrow heads, respectively. (**c**) Schematic models of interfacing β_{12} and χ_3 sheets (shaded red and blue, respectively) with mismatched and perpendicular boron rows. The large lattice mismatch causes high interfacial energies. (**d**) Structures of the β_{12} and χ_3 rows (top and middle, respectively) and an example of a new boron phase formed by assembling β_{12} and χ_3 rows (bottom). (**e**) Borophene sheet that contains domains with different periodic assemblies of β_{12} and χ_3 rows, including two new phases of borophene ($\nu_{7/36}$ and $\nu_{4/21}$ sheets). Here, $V_s = -1.3$ V. (Reproduced from Ref. [23]. (By courtesy of Boris I. Yakobson))

the unique borophene growth kinetics on Ag(111). Their theoretical calculations showed that the growth kinetics follow multiple energy pathways with close barriers along each edge but with strong anisotropy across the different edges. Therefore, the kinetic equilibrium shapes of borophene are a series of elongated hexagons, or nanoribbons.

3.3 Ag(110)

The successful growth of borophene on Ag(111) demonstrates that the interaction of boron and silver is appropriate for stabilization of 2D boron sheets without causing either clustering or alloying. Therefore, it is reasonable to assume that the other facets of silver crystals, e.g., Ag(110) and Ag(001), may also be good candidate materials for the growth of borophene. Because of the polymorphism of borophene, it is possible to realize other borophene phases when the symmetries and the lattice constants of the substrates change. In particular, the (110) surface is strongly anisotropic and is more reactive, which could favor the growth of 1D nanoribbons along the $[1\overline{1}0]$ direction. For example, Ag(110) and Au(110) surfaces have been used to grow graphene and silicene nanoribbons [27–29]. It is thus possible that borophene can also be used to form nanoribbons on Ag(110).

Based on these assumptions, Zhong et al. deposited boron on the Ag(110) surface at 570 K [30] using similar methods to those used for deposition on Ag(111). As expected, they obtained borophene nanoribbons (BNRs) that run along the $[1\overline{1}0]$ direction of Ag(110), as shown in Fig. 3.8. These BNRs were approximately 10 nm wide and could run as long as several hundred nanometers. Furthermore,

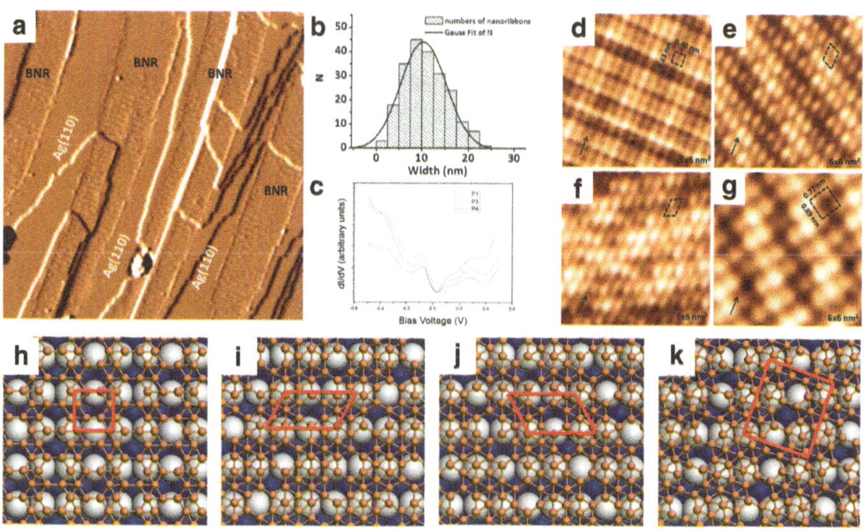

Fig. 3.8 (**a**) Derivative STM image showing boron nanoribbons grown on Ag(110). The image size is 100 × 100 nm^2. The nanoribbons run across the substrate steps without losing their continuity. (**b**) Systematic statistics for the nanoribbon width. A Gaussian fitting is shown as a black line. (**c**) The d*I*/d*V* spectra of borophene nanoribbons (BNRs) on Ag(110). The green, red, and blue curves represent the d*I*/d*V* spectra that were obtained on P1, P3, and P4, respectively. (**d**–**g**) High-resolution STM images of P1–P4, respectively. The unit cells of P1–P4 are marked using different dotted patterns. (**h**–**k**) Top views of the optimized P1–P4 BNRs on the Ag(110) surface, respectively. Color codes: B, small orange spheres; topmost Ag, large white spheres; lower Ag, large blue spheres. (Reproduced from Ref. [19])

the BNRs could even cross steps on the substrate without loss of continuity, which demonstrates the strong 2D characteristics of the BNRs.

To determine the atomic structures of these BNRs, Zhong et al. performed high-resolution STM measurements and found four different BNR phases. By combining their results with theoretical calculations, the atomic structures of the four BNRs were elucidated: the χ_3 sheet, the β sheets with two inequivalent directions, and the β_8 sheet. In fact, all these borophene polymorphs have previously been predicted to be stable in their freestanding form [15], which again proves the polymorphism of borophene. In particular, the χ_3 sheets can be synthesized on Ag(111) substrates and have been found to host anisotropic Dirac cones [22]. The restriction of massless Dirac fermions in a 1D nanoribbon could produce various exotic material properties. Therefore, the synthesis of χ_3 BNRs could stimulate the ongoing efforts to study the unusual properties that originate from quantum size effects.

3.4 Al(111)

Among the polymorphs of borophene, the honeycomb phase has the highest hole density ($\nu = 1/3$) and resembles the well-known graphene, which hosts massless Dirac fermions and numerous other exotic properties. Furthermore, the honeycomb boron structure exists naturally in the well-known high-T_c superconductor MgB_2 and is believed to be the origin of the superconductivity in MgB_2. Therefore, realization of honeycomb borophene could provide a good platform to investigate the interplay between the topological and superconducting orders. However, previous calculations show that the energy per boron atom is the highest of all the borophene polymorphs [31], thus indicating that freestanding honeycomb borophene is quite unstable.

One effective approach to increase the stability of honeycomb borophene is the use of electron doping, because the external charge will change the occupation of the bonding or antibonding orbitals and in turn will reorder the energies of all the polymorphs [32]. For honeycomb borophene, the energy per boron atom decreases continuously with increasing electron doping [32, 33]. Honeycomb borophene finally reaches the ground state when the doping reaches 0.5–1.0 e/atom. In other words, if we assume that each boron atom receives one electron, then the boron atom will become a "carbon" atom and the conventional 2c–2e σ bonds will dominate. Heavy electron doping of this type has been discovered in MgB_2, which consists of alternating honeycomb boron layers and triangular magnesium layers. It is accepted that the honeycomb boron layer contributes to the superconductivity, while the magnesium atoms act as the electron donors.

To achieve heavy electron doping of borophene, a suitable substrate will be required. According to the theoretical calculations for β_{12} and χ_3 borophenes on Ag(111), the charge transfer from Ag(111) to boron is quite small (<0.1 e/atom) and is thus insufficient to stabilize the honeycomb borophene. Recently, Li et al. used Al(111) as a substrate to grow borophene [34]. Because of the low work

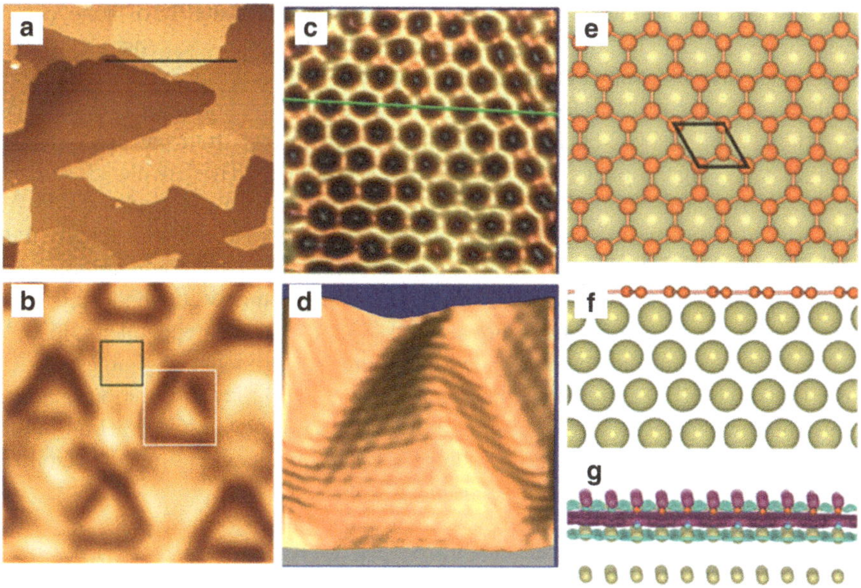

Fig. 3.9 (**a**) STM image (180 nm × 180 nm) of borophene islands on Al(111) surface. (**b**) STM image (15 nm × 15 nm) showing the large-period triangular corrugation. (**c**) High-resolution STM image (2.4 nm × 2.4 nm) of the area marked by the black rectangle in (**b**), showing a flat honeycomb lattice. (**d**) 3D STM image (4 nm × 4 nm) of the area marked by the white rectangle in (**b**). (**e–g**) Structural models and the electron density of honeycomb borophene on Al(111) and Ag(111) surfaces. (**e**) and (**f**) Top and side views of honeycomb borophene on Al(111), respectively. (**g**) Electron density map of honeycomb borophene on Al(111). (Reproduced from Ref. [34])

function of aluminum, the electron doping of boron was as high as 0.7 e/atom, which made it possible to realize honeycomb borophene. The substrate temperature was maintained at 500 K during the growth process. Using STM, they observed borophene islands with a height of 320 pm, which corresponds to the height of a single-layer borophene. The measured height varied little when the bias voltage was varied between −6 V and + 6 V. Their high-resolution STM images of borophene showed a perfect honeycomb lattice (Fig. 3.9c), and the lattice constant (0.29 nm) was very close to the calculated results for both freestanding honeycomb borophene (0.3 nm) and the Al(111) surface (0.286 nm). However, the small lattice mismatch also induces strain in borophene, which produced a periodic corrugation pattern in the STM images. Their theoretical calculations of the borophene/Al(111) structure confirmed the honeycomb lattice nature of the borophene. In addition, the honeycomb borophene remained flat on unreconstructed Al(111), similar to the case of borophene on Ag(111). In addition to the large charge transfer from Al(111), the nearly perfect lattice matching between honeycomb borophene and the Al(111) surface provides another reason for the stabilization of the honeycomb borophene.

To determine the mechanism underlying the enhanced stability of honeycomb borophene on Al(111), Zhu et al. investigated the structural, energetic, electronic, and lattice dynamic properties of this structure via first-principles calculations [35]. They found that the lowest energy configuration for honeycomb borophene on Al(111) is the FCC-HCP configuration, which differs from the predominant TOP-FCC configuration of graphene on metal surfaces. The calculated vertical spacing between the borophene and the Al(111) surface was as small as 1.54 Å, which is also much smaller than that in the graphene case. Most strikingly, the adhesion energy of the honeycomb borophene in the FCC-HCP configuration is as high as 1.05 eV/atom, which indicates that the interaction between the honeycomb borophene and the substrate is not a weak van der Waals-type interaction. A Bader charge density analysis revealed that there was pronounced charge transfer of approximately 1.0 e per B atom from the top Al(111) layer to the borophene. This significant charge transfer, along with the covalent bonding interaction in the interface region, resulted in a strong adhesion energy that could ultimately stabilize the Al(111)-supported honeycomb borophene.

3.5 Au(111)

Gold is another noble metal and has lower chemical reactivity than silver. By depositing boron atoms on the Au(111) surface at 550 °C, Kiraly et al. [36] found that the herringbone reconstruction of the Au(111) transformed into a trigonal network to relieve the strain, as shown in Fig. 3.10a. Interestingly, small nanoscale borophene islands emerged at the nodes of the trigonal network, as indicated by the white-dashed lines in Fig. 3.10c. Theoretical calculations show that these borophene islands are $\nu_{1/12}$ sheets [16]. The trigonal network shows a characteristic periodicity within the 5.5 to 8.0 nm range and acts as a template for nanoscale borophene island growth. Increasing the boron coverage led to the breakdown of the trigonal network and formation of larger borophene islands, but these larger borophene islands were embedded in the topmost Au(111) layer. As the boron dose increased, the borophene sheets increased in size to produce domains that extended for at least several tens of nanometers.

During the growth of boron on Au(111), the total atomic boron dose required to observe the nanoscale borophene islands is an order of magnitude larger than that required for growth on Ag(111), which indicates that the boron may dissolve into the Au(111) bulk. When boron is deposited at room temperature, it will form clusters on the Au(111) surface, and the XPS spectra show a clear peak in the B 1 s core level (Fig. 3.11a), which is expected as a result of the limited solubility of boron in Au at room temperature. However, when the substrate temperature reaches 550 °C, the B 1s peak is significantly reduced, thus indicating that the boron surface coverage is much lower in these cases. This scenario has also been confirmed by time-of-flight secondary ion mass spectrometry (ToF-SIMS), as illustrated in Fig. 3.11b. The boron signals for the boron cluster sample are higher in the SIMS spectra when

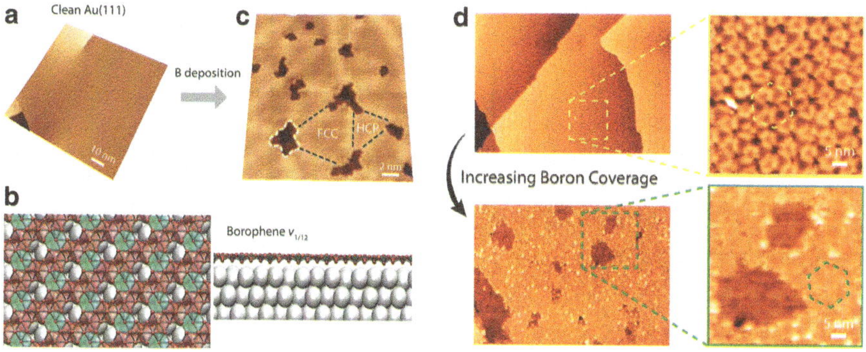

Fig. 3.10 (**a**) STM image of a clean Au(111) surface that shows a herringbone reconstruction ($V = -1.5$ V, $I = 100$ pA). (**b**) Following boron deposition at 550 °C, the herringbone reconstruction is modified to form a trigonal network, where nanoscale borophene islands (one is highlighted using a white-dashed line) emerge at the nodes and result in templated growth across the surface ($V = -0.1$ V, $I = 200$ pA). (**c**) Atomic structure of borophene $v_{1/12}$ that was computationally modeled and predicted for Au(111) (left, top-down view; right, cross-sectional view). (**d**) Increasing boron dose results in breakdown of the trigonal network and growth of larger borophene islands (top panels, $V = -2$ V, $I = 100$ pA; bottom panels, $V = 3.4$ V, I = 60 pA). (Reproduced from Ref. [36]. (By courtesy of Boris I. Yakobson))

compared with those of the trigonal network sample, thus indicating a higher surface concentration of boron. This observation agrees well with the XPS B 1s level results. Depth profiling of the four samples revealed that the boron concentration in the trigonal network sample only decays after a depth of 30 nm into the bulk, whereas the boron signals decayed more rapidly in the three other cases after the surface was sputtered. Therefore, during borophene growth on Au(111), the boron atoms will dissolve into the Au(111) bulk at high temperature and are then segregated to the surface to form borophene when the sample cools down, as shown schematically in Fig. 3.11c. This growth mode is similar to that of graphene on a metal substrate such as Cu(111).

To provide further understanding of the growth mode mechanisms of boron on Au(111), Kiraly et al. [36] performed density functional theory calculations to quantify the adsorption energetics of the boron atoms. Their calculation results indicated that while both the bulk solutions and the surface locations are energetically favorable, they are both significantly higher in energy than the near-surface dissolution of the boron atoms. In particular, the most stable configurations composed of isolated boron atoms are located in subsurface sites in the topmost layer, with sites along the Au(111) step edges being favored by 0.3 eV over the planar subsurface sites. These energetics differ significantly from those of boron on the Ag(111) substrate. Additionally, this behavior marks a significant difference from the 2D surface growth motif of the borophene overlayers on Ag(111), thus confirming the experimental observations.

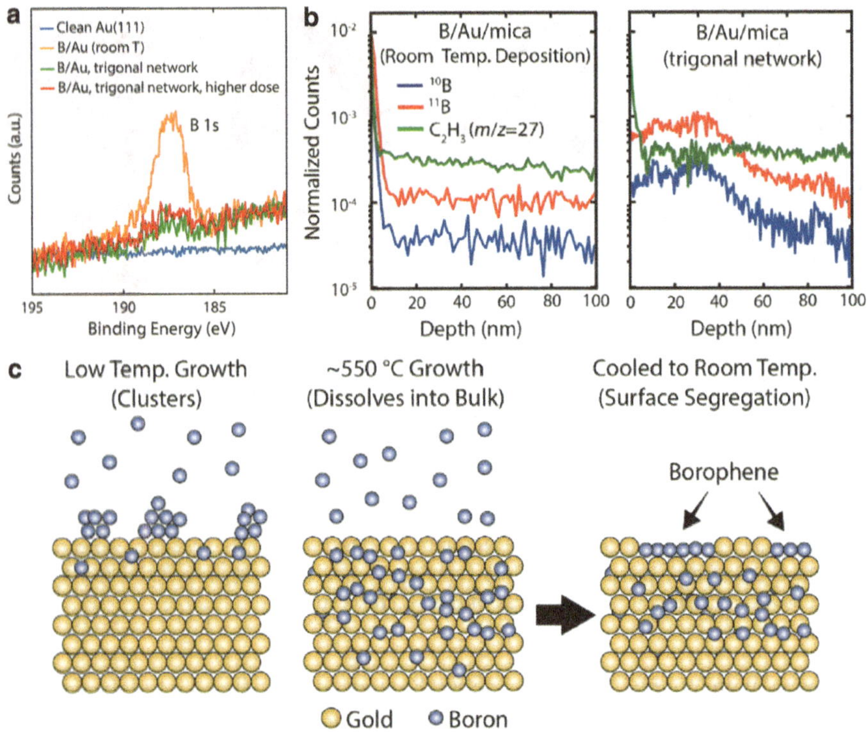

Fig. 3.11 (**a**) B 1s core-level spectra for room-temperature B deposition (orange), trigonal network with low dose (green), trigonal network with high dose (red), and clean Au(111) (blue). (**b**) SIMS spectra showing that boron is limited to the surface when deposited at room temperature but penetrates 30 nm into the subsurface region in the trigonal network sample that was prepared at high temperature. (**c**) Schematic illustration of borophene growth dynamics. At a low substrate temperature, the boron is predominantly located on the surface and forms boron clusters. For higher substrate temperatures, the boron dissolves into the bulk and is then segregated to the surface upon cooling to form 2D borophene sheets. (Reproduced from Ref. [36]. (By courtesy of Boris I. Yakobson))

3.6 Cu(111)

Despite the promise of the material's unique chemical and physical properties, borophene still faces the critical issue that the sizes of its domains are limited. When borophene is grown on Ag(111), Al(111), or Au(111) surfaces, the borophene domains range in size from several to hundreds of nanometers, which is far below the limits required for device fabrication. Therefore, there is a strong demand for synthesis of large-scale borophene domains that are feasible for use in device applications.

To produce larger domains, it is desirable to explore candidate substrates that are less inert than Ag, i.e., substrates that are potentially able to promote the growth of larger domains, but not so reactive as to form metal-boride compounds. With this aim, Wu et al. used Cu(111), which was a more reactive substrate relative to Ag(111) and Au(111) [14]. The substrate temperature was maintained at 770 K during growth. They used low-energy electron microscopy (LEEM) and LEED to perform simultaneous real-time growth monitoring and *k*-space structural characterization. With increasing boron coverage, the borophene/Cu(111) domain size can reach 10–100 μm^2, and some of the borophene flakes even crossed several of the step edges of the substrate, as shown in Fig. 3.12(a). The borophene/Cu(111) atomic structure was further characterized using STM with a CO-functionalized tip. Combining the LEED and STM results with theoretical calculations, they proposed that the borophene on Cu(111) structure would be composed of a triangular network with an HH concentration of $\nu = 1/5$, i.e., the same as that for the χ_3 borophene, as illustrated in Fig. 3.12b. In addition, their first-principles calculations showed that there was no significant covalent bonding between borophene and Cu(111), despite the occurrence of charge transfer from the Cu(111) surface to borophene.

Detailed studies of the growth process of borophene on Cu(111) show that it is a self-limiting process. This is advantageous for the synthesis of heterostructures based on atomically thin layers. Indeed, beyond the 1 ML coverage, the growth rates on Cu and Ag both decrease dramatically, even if the boron flux is substantially increased.

The successful synthesis of borophene on Cu(111) makes future borophene transfer and device applications possible, because Cu can easily be dissolved into chemical solutions such as $FeCl_3$. However, Cu(111) single crystals are usually small and expensive, making them unsuitable for the study of either transport properties or device applications. Wu et al. addressed this issue by growing borophene on nanometer-thick Cu(111) films on sapphire [37]. They grew high-quality Cu(111) thin films on a sapphire substrate by the secondary-grain growth method, as follows: an amorphous copper film is first grown on the sapphire surface and is then crystallized in situ by post-annealing at an elevated temperature. The borophene layer that they obtained can cover the entire copper film surface, which may stimulate further experimental efforts to fabricate borophene devices by transferring borophene onto other substrates.

Recently, Yue et al. performed independent experiments by depositing boron on a Cu(111) surface and obtained samples with an unexpected herringbone-like structure [38]. Their combination of STM measurements and DFT calculations showed that boron and copper formed a 2D copper boride: Cu_8B_{14}. Furthermore, they also found that this copper boride forms a single phase with metallic conductivity, akin to borophene grown on Cu(111). These results challenge the formation of borophene on Cu(111).

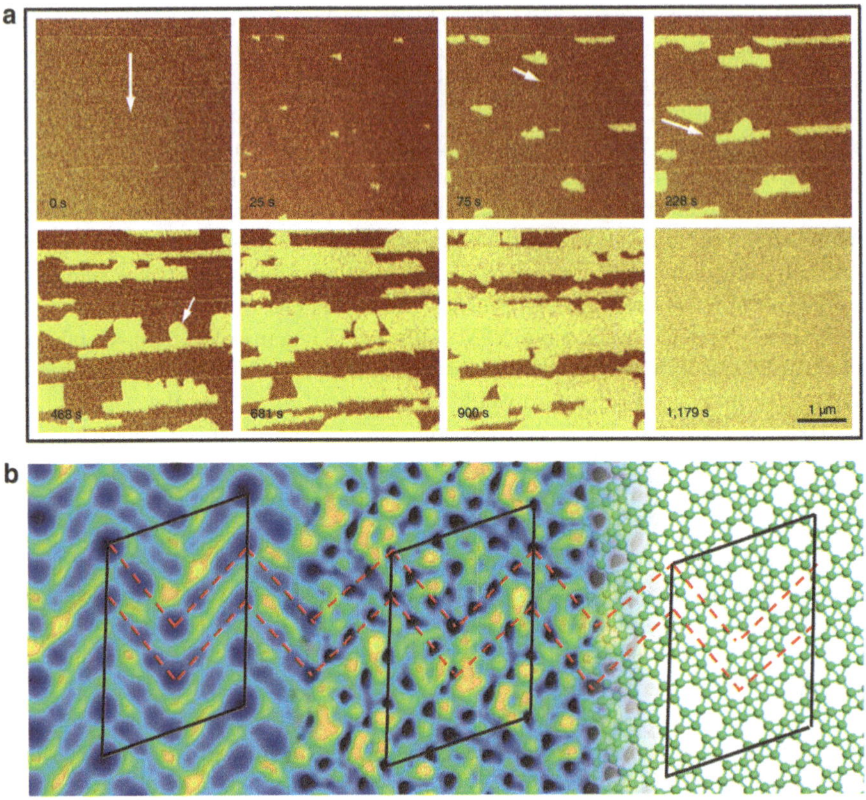

Fig. 3.12 (**a**) Sequence of bright-field LEEM images acquired at $T = 770$ K, which reveals that the borophene islands nucleate preferentially from the down-step edges of the Cu terraces (i.e., in the direction marked by the arrow in the first panel at $t = 0$ s); additionally, the growth proceeds more quickly along the step-edge direction at 97 $\text{Å}\text{s}^{-1}$ compared with 20 $\text{Å}\text{s}^{-1}$ in the perpendicular direction. These two panels, along with the image at 468 s, also show that the down-step nucleation is followed by borophene growth across the upper-step edge. In contrast to the down-step growth, the up-step growth generates arc-shaped islands, indicating more isotropic growth kinetics (see the arrow in the 468 s panel). A continuous monolayer (ML) emerges that reproduces the Cu(111) terrace structure faithfully. The growth rate was ~0.05 ML min^{-1}. (**b**) Left: ultrahigh-resolution STM data for borophene (set-point parameters: $I_s = 50$ pA and $V_b = +20$ mV). The rhomboid corresponds to the borophene unit cell, and the red-dashed lines serve as visual guides along the zigzag pattern characteristic of all STM data. Middle: density functional theory (DFT)-simulated constant tunneling current isosurface of the proposed borophene structure while assuming a p_z tunneling state for the tip. Right: pictorial view of the borophene structure with the boron atoms and bonds shown in green. The red-dashed lines continue the basic zigzag motif from the left panel obtained from the DFT simulations (middle panel) and the proposed ball-and-stick structure. (Reproduced from Ref. [14]. (By courtesy of Adrian Gozar))

3.7 Ir(111)

Another suitable substrate for borophene growth is Ir(111). When submonolayer boron atoms were deposited on Ir(111) at substrate temperatures ranging from ~300 °C to 600 °C, a single borophene phase formed on the surface [39]. Figure 3.13(a, b) show the LEED patterns of pristine Ir(111) and borophene on Ir(111), respectively. A detailed structural analysis showed that this borophene phase could be described as a (6 × 2) superlattice with respect to the 1 × 1 lattice of Ir(111). Figure 3.13c shows a large-scale STM image of the borophene/Ir(111) structure, where approximately 80% of the surface was uniformly covered with a striped, complex-structured planar overlayer. There are three 120° rotational

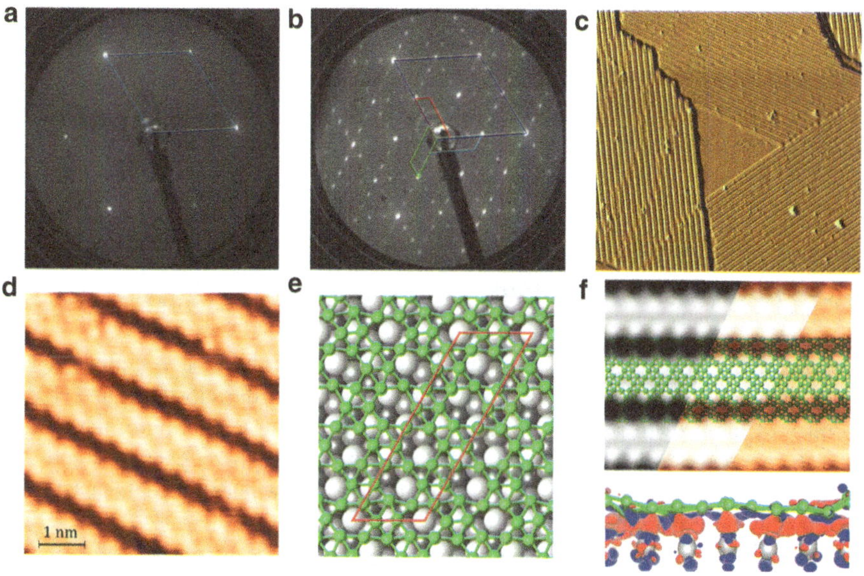

Fig. 3.13 (**a**) LEED pattern from clean Ir(111); $E = 85$ eV. (**b**) LEED pattern from borophene/Ir(111); $E = 85$ eV. Reciprocal unit cells of Ir and borophene are sketched in (**a**) and (**b**), respectively, while three equivalent orientations of borophene are indicated in (**b**) as the red, light blue, and green reciprocal unit cells. (**c**) Comparison between the size and the geometry of the Ir (blue) and borophene (red) unit cells on the (111) Ir face. (**d**) High-resolution STM image of borophene ($V_{bias} = 1$ V, $I = 500$ pA). (**e**) Top view of χ_6 borophene on Ir(111), where the boron atoms are green. The unit cell is enclosed within the red line. (**f**) Top panel – Left: DFT-simulated constant tunneling current isosurface of the χ_6 borophene structure in (**a**). Integration energy range [0, +1] eV, isostate density = 0.0023. Right: a cut from raw experimental data, with $V_b = 1$ V and $I_t = 350$ pA. In the center of the panel, a proposed atomic arrangement from (a) is overlaid. Bottom panel: a (100) view of the isosurface plot (0.065 e Å^{-3}) of the charge redistribution between Ir and borophene. The excess electron density (negative charge) is shown in red, and the corresponding lack of electron density (positive charge) is shown in blue. (Reproduced from Ref. [39]. Reprinted by permission from Vinogradov et al. 2019. Copyright (2019) American Chemical Society)

domains because of the sixfold symmetry of the substrate. The stripes are oriented along the equivalent [101], [110], and [011] directions of Ir(111).

The magnified STM image in Fig. 3.13d shows that each stripe consists of two bright undulating curves that are matching in registry. The periodicity of the stripes was determined to be approximately 1.45 nm, and the period of the undulation along an individual stripe was approximately 0.55 nm. The proposed structural model for borophene on Ir(111) is χ_6 borophene, which is shown in Fig. 3.13e. The stability of χ_6 borophene on Ir(111) was confirmed by DFT calculations. Additionally, the simulated STM images agree well with the experimental results, as illustrated in the top panel of Fig. 3.13f.

The charge transfer between borophene and Ir(111) can be studied based on a Bader analysis, as shown in Fig. 3.13d. Interestingly, the calculation results show that the χ_6 borophene donates approximately two electrons per unit cell to the metal substrate. Therefore, χ_6 borophene is hole-doped on Ir(111); this is in stark contrast to borophene on Cu(111), where the metal substrate donates electrons to the borophene. This difference in electron doping for χ-borophene on Ir(111) and on Cu(111) originates from their different HH densities of $\nu = 1/6$ and $\nu = 1/5$, respectively. This behavior correlates with the Pauling electronegativity scale, which indicates that borophene should donate electrons to Ir and attract electrons from Cu.

3.8 Summary and Outlook

In this chapter, we see that various borophene polymorphs have been synthesized by MBE in an ultrahigh vacuum. By appropriate selection of the substrates and careful control of the growth parameters, including the substrate temperature, the B flux, and the coverage, it is possible to synthesize certain borophene phases selectively. However, it is still difficult to synthesize large-scale single-phase borophene because of the polymorphism of borophene.

To date, borophenes can only be synthesized on metallic substrates, including Ag, Au, and Al surfaces. The metallic substrate enables investigation of the structures and the electronic properties of borophenes using spectroscopic techniques, such as STM, LEED, XPS, and angle-resolved photoemission spectroscopy (ARPES). Experimental works on the electronic structures of borophenes will be discussed in the next chapter. However, from a practical viewpoint, the electrical shorting of the metallic substrates hampers the transport measurements, which makes it impossible to fabricate borophene devices on these metallic substrates. Therefore, it is highly desirable to transfer the borophenes from metallic to insulating substrates or fabricate borophenes directly on insulating substrates.

References

1. A.R. Organov, V.L. Solozhenko, Boron: A hunt for superhard polymorphs. J. Superhard Mater. **31**, 285 (2009)
2. B. Albert, H. Hillebrecht, Boron: Elementary challenge for experimenters and theoreticians. Angew. Chem. Int. Ed. **48**, 8640–8668 (2009)
3. Z. Zhang, E.S. Penev, B.I. Yakobson, Chem. Soc. Rev. **46**, 6747 (2017)
4. L. Kong, K. Wu, L. Chen, Front. Phys. **13**, 138105 (2018)
5. A. Molle, J. Goldberger, M. Houssa, Y. Xu, S.-C. Zhang, D. Akinwande, Buckled two-dimensional Xene sheets. Nat. Mater. **16**, 163 (2017)
6. J. Zhao, H. Liu, Z. Yu, R. Quhe, S. Zhou, Y. Wang, C.C. Liu, H. Zhong, N. Han, J. Lu, Y. Yao, K. Wu, Rise of silicene: A competitive 2D material. Prog. Mater. Sci. **83**, 24–151 (2016)
7. Y. Liu, E.S. Penev, B.I. Yakobson, Probing the synthesis of two-dimensional boron by first-principles computations. Angew. Chem. Int. Ed. **52**, 3156–3159 (2013)
8. L.Z. Zhang, Q.B. Yan, S.X. Du, G. Su, H.-J. Gao, Boron sheet adsorbed on metal surfaces: Structures and electronic properties. J. Phys. Chem. C **116**, 18202–18206 (2012)
9. H. Liu, J. Gao, J. Zhao, From boron cluster to two-dimensional boron sheet on cu(111) surface: Growth mechanism and hole formation. Sci. Rep. **3**, 3238 (2013)
10. A.J. Mannix, X.-F. Zhou, B. Kiraly, J.D. Wood, D. Alducin, B.D. Myers, X. Liu, B.L. Fisher, U. Santiago, J.R. Guest, M.J. Yacaman, A. Ponce, A.R. Oganov, M.C. Hersam, N.P. Guisinger, Science **350**, 1513 (2015)
11. B. Feng, J. Zhang, Q. Zhong, W. Li, S. Li, H. Li, P. Cheng, S. Meng, L. Chen, K. Wu, Nat. Chem. **8**, 563 (2016)
12. G.P. Campbell, A.J. Mannix, D.J. Emery, T.-L. Lee, N.P. Guisinger, M.C. Hersam, M.J. Bedzyk, Nano Lett. **18**, 2816 (2018)
13. X. Liu, L. Wang, S. Li, M.S. Rahn, B.I. Yakobson, M.C. Hersam, Nat. Commun. **10**, 1642 (2019)
14. R. Wu, I.K. Drozdov, S. Eltinge, P. Zahl, S. Ismail-Beigi, I. Bozovic, A. Gozar, Nat. Nanotechnol. **14**, 44–49 (2019)
15. X. Wu, J. Dai, Y. Zhao, Z. Zhuo, J. Yang, X.C. Zeng, Two-dimensional boron monolayer sheets. ACS Nano **6**, 7443–7453 (2012)
16. Z. Zhang, Y. Yang, G. Gao, B.I. Yakobson, Angew. Chem., Int. Ed. **54**, 13022–13026 (2015)
17. S. Xu, Y. Zhao, J. Liao, X. Yang, H. Xu, The nucleation and growth of borophene on the Ag(111) surface. Nano Res. **9**, 2616–2622 (2016)
18. B. Peng et al., Stability and strength of atomically thin borophene from first principles calculations. Mater. Res. Lett. **5**, 399 (2017)
19. Q. Zhong, J. Zhang, P. Cheng, B. Feng, W. Li, S. Sheng, H. Li, S. Meng, L. Chen, K. Wu, J. Phys. Condens. Matter **29**, 095002 (2017)
20. B. Feng, J. Zhang, R.-Y. Liu, T. Iimori, C. Lian, H. Li, L. Chen, K. Wu, S. Meng, F. Komori, I. Matsuda, Phys. Rev **B 94**, 041408(R) (2016)
21. B. Feng, O. Sugino, R.-Y. Liu, J. Zhang, R. Yukawa, M. Kawamura, T. Iimori, H. Kim, Y. Hasegawa, H. Li, L. Chen, K. Wu, H. Kumigashira, F. Komori, T.-C. Chiang, S. Meng, I. Matsuda, Phys. Rev. Lett. **118**, 096401 (2017)
22. B. Feng, J. Zhang, S. Ito, M. Arita, C. Cheng, L. Chen, K. Wu, F. Komori, O. Sugino, K. Miyamoto, T. Okuda, S. Meng, I. Matsuda, Adv. Mater. **30**, 1704025 (2018)
23. X. Liu, Z. Zhang, L. Wang, B.I. Yakobson, M.C. Hersam, Nat. Mater. **17**, 783–788 (2018)
24. H. Tang, S. Ismail-Beigi, Phys. Rev. Lett. **99**, 115501 (2007)
25. X. Liu, Z. Wei, I. Balla, A.J. Mannix, N.P. Guisinger, E. Luijten, M.C. Hersam, Sci. Adv. **3**, e1602356 (2017)
26. Z. Zhang, A.J. Mannix, X. Liu, Z. Hu, N.P. Guisinger, M.C. Hersam, B.I. Yakobson, Sci. Adv. **5**, eaax0246 (2019)
27. L. Massimi, O. Ourdjini, E. Cavaliere, L. Gavioli, A.D. Pia, C. Mariani, M.G. Betti, AIP Conf. Proc. **1667**, 020002 (2015)

28. B. Aufray, A. Kara, S. Vizzini, H. Oughaddou, C. Leandri, B. Ealet, G. Le Lay, Appl. Phys. Lett. **96**, 183102 (2010)
29. B. Feng, H. Li, S. Meng, L. Chen, K. Wu, Surf. Sci. **645**, 74–79 (2016)
30. Q. Zhong, L. Kong, J. Gou, W. Li, S. Sheng, S. Yang, P. Cheng, H. Li, K. Wu, L. Chen, Phys. Rev. Mater. **1**, 021001 (2017)
31. E.S. Penev, S. Bhowmick, A. Sadrzadeh, B.I. Yakobson, Nano Lett. **12**, 2441–2445 (2012)
32. Z. Zhang, S.N. Shirodkar, Y. Yang, B.I. Yakobson, Angew. Chem. Int. Ed. **56**, 15421–15426 (2017)
33. S.N. Shirodkar, E.S. Penev, B.I. Yakobson, Sci. Bull. **63**, 270–271 (2018)
34. W. Li, L. Kong, C. Chen, J. Gou, S. Sheng, W. Zhang, H. Li, L. Chen, P. Cheng, K. Wu, Sci. Bull. **63**, 282 (2018)
35. L. Zhu, B. Zhao, T. Zhang, G. Chen, S.A. Yang, J. Phys. Chem. C **123**, 14858–14864 (2019)
36. B. Kiraly, X. Liu, L. Wang, Z. Zhang, A.J. Mannix, B.L. Fisher, B.I. Yakobson, M.C. Hersam, N.P. Guisinger, ACS Nano **13**, 3816–3822 (2019)
37. R. Wu, A. Gozar, I. Bozovic, npj Quantum Mater. **4**, 40 (2019)
38. C. Yue, X.-J. Wang, G. Gao, A. R. Oganov, X. Dong, X. Shao, X. Wang, J. Sun, B. Xu, H.-T. Wang, X.-F. Zhou, and Y. Tian, arXiv:1912.06027 (2019)
39. N.A. Vinogradov, A. Lyalin, T. Taketsugu, A.S. Vinogradov, A. Preobrajenski, ACS Nano. **13**, 14511 (2019)

Chapter 4
Electronic Structures of Borophene

Baojie Feng, Osamu Sugino, and Kehui Wu

4.1 Metallicity

Located between metallic beryllium and nonmetallic carbon, boron has only three valence electrons. In its bulk form, boron has a large family of crystalline polymorphs, and most of these polymorphs have icosahedral B_{12} cages as their fundamental building block [1, 2]. Because of this special structure, almost all the bulk boron crystals are insulators or semiconductors at ambient pressure. In the two-dimensional (2D) limit, most borophene polymorphs share a triangular lattice with periodic hollow hexagons. This unique material structure favors multicenter bonds, which makes most of the borophene polymorphs metallic [3–6]. It should be noted here that 2D metallic materials are quite rare, despite the recent upsurge in the study of 2D materials. For example, graphene is a semimetal with a vanishing density of states near the Fermi level; phosphorene is a semiconductor with a moderate band gap; and single-layer hexagonal boron nitride is a complete insulator. From this viewpoint, metallic borophenes could serve as ideal compensation structures for other 2D materials and could be used as ultrathin electrodes or other conductive parts in future quantum devices.

The recent synthesis of borophenes makes it possible to verify the metallicity of these materials [7, 8]. To date, several borophene polymorphs have been synthesized successfully on noble metal substrates via molecular beam epitaxy, including the β_{12} phase [8, 9], the χ_3 phase [8], the α phase [10], and the honeycomb phase [11]. Because of the presence of the metallic substrates, traditional transport

B. Feng · K. Wu (✉)
Institute of Physics, Chinese Academy of Sciences, Beijing, China
e-mail: khwu@iphy.ac.cn

O. Sugino
The Institute for Solid State Physics, The University of Tokyo, Kashiwa, Chiba, Japan

I. Matsuda, K. Wu (eds.), *2D Boron: Boraphene, Borophene, Boronene*,
https://doi.org/10.1007/978-3-030-49999-0_4

measurements were quite difficult to perform because of the electrical shorting properties of the substrates. In addition, the oxidation of the borophenes [8], although weak, also hampers measurement of the intrinsic properties of borophene via ex situ measurements.

The first experimental evidence of the metallic character of borophene was obtained by in situ scanning tunneling spectroscopy (STS) measurements [7, 8]. STS can detect the local density of states (LDOS) near the Fermi level and can thus provide evidence for the existence or absence of an energy gap at the Fermi level. Typical STS spectra for β_{12} and χ_3 borophenes are shown in Fig. 4.1(a, b). The spectra show that there is still a considerable LDOS around the Fermi level in both borophenes, which is a signature of the metallic character of these materials. These results show good qualitative agreement with the calculated LDOS, as illustrated in Fig. 4.1(c–f).

However, given that both the β_{12} and χ_3 borophenes were grown on Ag(111), it is possible that the electrons tunnel through the borophene from the STM tip to the substrates, or vice versa. Therefore, the measured STS signal may originate from the electronic states of the tip or the substrate and not from the borophenes. Furthermore, the possible band hybridization of the borophene and the Ag(111) substrate makes it difficult to untangle the contribution from the borophene and the substrate. These facts indicate that the STS results do not provide conclusive evidence for the metallic character of borophene. Another more powerful technique that can be used to investigate the electronic structures of materials is angle-resolved

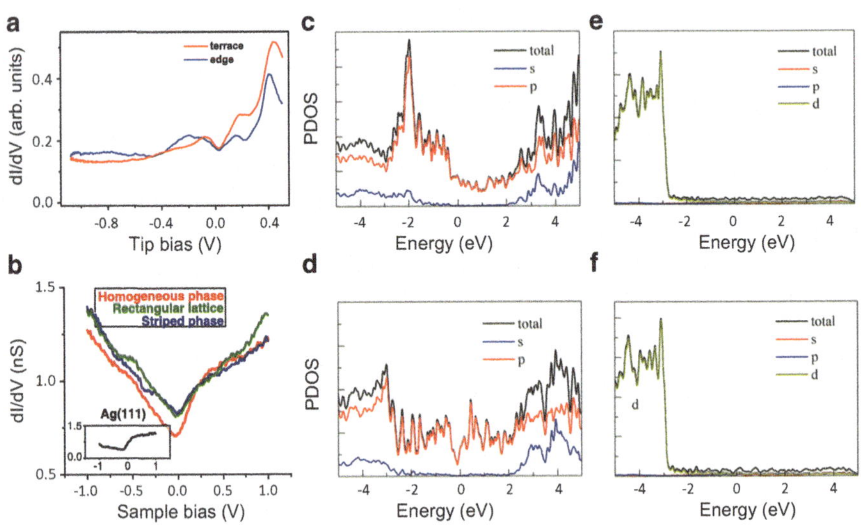

Fig. 4.1 (**a, b**) Scanning tunneling spectroscopy measurements of β_{12} and χ_3 borophenes. (**c–f**) Calculated densities of states for the β_{12} and χ_3 borophenes and the Ag(111) substrates. (Reproduced from Ref. [7, 8]. Reprinted by permission from Mannix et al. 2015. Copyright (2015) American Association for the Advancement of Science)

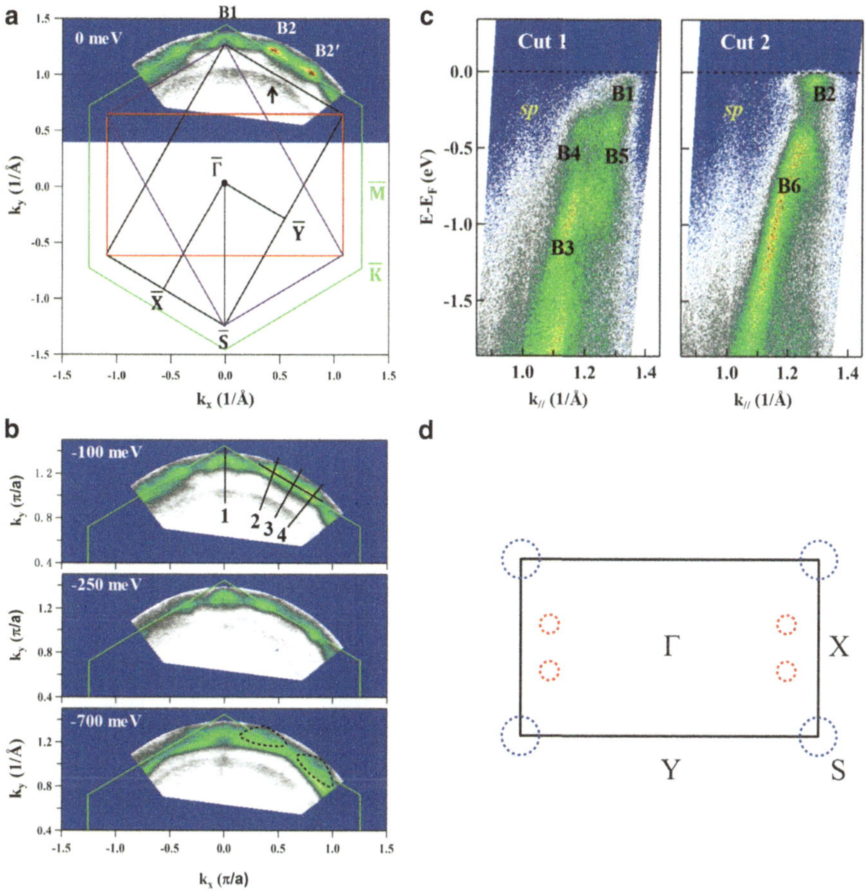

Fig. 4.2 (**a**) ARPES intensity plot of Fermi surface of β_{12} borophene on Ag(111). The black solid arrow indicates the bulk *sp* band of Ag(111). The green hexagon indicates the Brillouin zone (BZ) of Ag(111); the black, purple, and red rectangles indicate the BZs of monolayer boron for the three equivalent directions. (**b**) Constant energy contours at different binding energies. (**c**) ARPES intensity plots along Cut 1 and Cut 2, as indicated by the black lines shown in (**b**). (**d**) Schematic drawing of the Fermi surfaces with the Brillouin zone of the boron sheet. (Reproduced from Ref. [12])

photoemission spectroscopy (ARPES). By combining ARPES with first-principles calculations, it is possible to distinguish the electronic band structure of the substrate and that of the borophene.

Figure 4.2(a, b) show the ARPES spectral weight of the Fermi surface and the constant energy contours (CECs) of the β_{12} borophene on Ag(111) [12]. The Fermi surface consists of pocket-like bands denoted by B1, B2, and B2$'$. These bands can be better resolved from the band structures along specific momentum cuts, as shown in Fig. 4.2(c). From the ARPES results, it can be summarized that several

Fermi pockets exist in the Brillouin zone, where one electron pocket is centered at the S point and a pair of electron pockets exist inside the *X* point, as indicated by the blue and red circles shown in Fig. 4.2(d), respectively. In addition to the Fermi pockets, the remnant signals of the Ag(111) bulk *sp* band can also be identified, as indicated in Fig. 4.2(a, c). Because the Ag(111) substrate bands are well separated from the electron pockets of the borophene and maintain their original dispersion, any hybridization of the boron sheet with Ag(111) appears to be weak, which agrees well with the previous theoretical and experimental results [8, 13, 14]. Therefore, the ARPES results provide compelling evidence for the metallic character of borophene. Subsequently, the metallic character of the χ_3 borophene was also confirmed via ARPES measurements [15].

The experimental works on β_{12} and χ_3 borophene have stimulated further theoretical studies of the intriguing metallic behavior of borophene. Recently, Zhang et al. reported that the intrinsic resistivity of borophene can be tuned easily by simply adjusting the carrier densities, in contrast to the behavior of graphene and conventional metals [16]. Liu et al. reported that the intrinsic resistivity of both β_{12} and χ_3 borophene shows a nontrivial anisotropy that amounts to at least 28% [17]. In addition, the two borophenes have complex Fermi surfaces with multiple branches that lead to the breakdown of the Bloch-Grüneisen theory. These works indicate that monolayer boron can serve as an intriguing platform for realization of tunable and anisotropic two-dimensional electronic devices.

4.2 Dirac Cones

In a wide range of materials, i.e., Dirac materials, low-energy excitations behave as massless Dirac fermions rather than as conventional fermions that obey the Schrodinger equation [18–20]. One prototypical feature in the electronic structures of Dirac materials is the existence of gapless Dirac cones (Fig. 4.3a) with unique spin or pseudospin characteristics that generate exotic properties, including high carrier mobility, the quantum spin Hall effect, and extremely large magnetoresistance. As a result, Dirac materials have attracted major attention in recent years, and the search for new Dirac materials is at the forefront of materials science research.

Because of the polymorphism of borophene, the possibility that some borophene phases will host Dirac cones has been studied intensively. The first borophene that was found to host Dirac cones was the 8-Pmmn borophene that was predicted theoretically by Zhou et al. in 2014 [21]. Further theoretical studies have shown that the existence of these Dirac cones endows 8-Pmmn borophene with various exotic properties [22–25], which makes it a promising material for use in device applications. Subsequently, several other borophene phases have been predicted to host Dirac cones, including the P6/mmm phase [26] and the honeycomb phase [27]. However, experimental synthesis of these borophene phases remains a major challenge and will require further experimental effort.

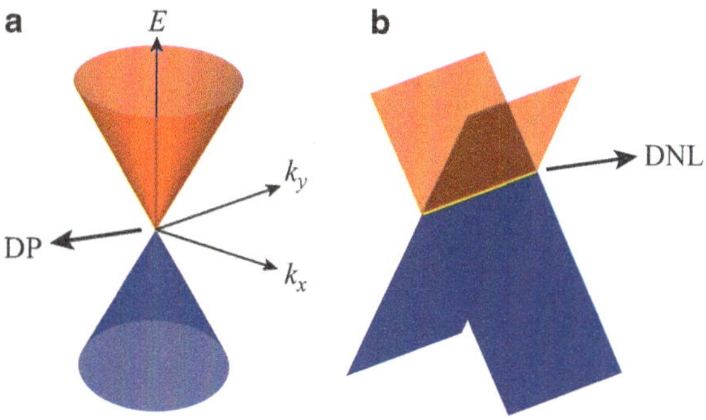

Fig. 4.3 Schematic drawings of (**a**) the Dirac cone and (**b**) the Dirac nodal line. The Dirac point (DP) and the Dirac nodal line (DNL) are indicated by a yellow dot and a yellow line, respectively

The first borophene that was experimentally confirmed to host Dirac cones was the β_{12} phase [28]. Feng et al. performed in situ ARPES measurements on β_{12} borophene that was grown on Ag(111) and discovered a pair of Dirac cones inside the X point, with Dirac points being located 0.25 eV below the Fermi level. Linear dichroism ARPES measurements show that the Dirac bands are only made visible using p-polarized photons, as shown in Fig. 4.4(g). A simple analysis based on photoemission matrix element effects shows that the Dirac bands are mainly derived from the p_z orbitals of boron, which was also the result of the first-principles calculation. The calculation was analyzed to derive a tight-binding (TB) model mainly comprised of the p_z orbitals of boron. The TB model shows that the appearance of Dirac cone is due to the existence of two equivalent sublattices A and B. The amplitude of wave function on one sublattice (A) is either 1, $\exp[2\pi i/3]$, or $\exp[4\pi i/3]$, while the amplitude on the other sublattice (B) vanishes because the amplitudes of A destructively interfere on B as $1 + \exp[2\pi i/3] + \exp[4\pi i/3] = 0$. The destructive interference occurs on β_{12} lattice as on honeycomb lattice, although the former has additional site c at the center of the hexagon, where the amplitude is vanishing as in the sublattice B as illustrated in Fig. 4.4(a–c). Therefore, like graphene, Dirac cone appears in the Brillouin zone equivalent to the K point of the honeycomb lattice on the β_{12} structure. Note that the Dirac cones at the K points of the honeycomb lattice will be folded into the Brillouin zone of the β_{12} borophene, as shown in Fig. 4.4(d).

The TB model above reveals the existence of one Dirac cone inside each X point (Fig. 4.4d, e), while the ARPES measurements show a pair of cones. The reason for this is that the TB analysis above is based on freestanding β_{12} borophene. When the β_{12} borophene is grown on Ag(111), striped moiré patterns emerge because of the lattice mismatch between the borophene and the substrate. The effects of the moiré pattern can be treated as external periodic perturbations in the TB model. The modified TB model shows that the Dirac cones will be split into a pair because of the breaking of the sublattice symmetry, which agrees well with the experimental results. Similar substrate-induced splitting of Dirac cones has also been observed in silicene on Ag(111) [29].

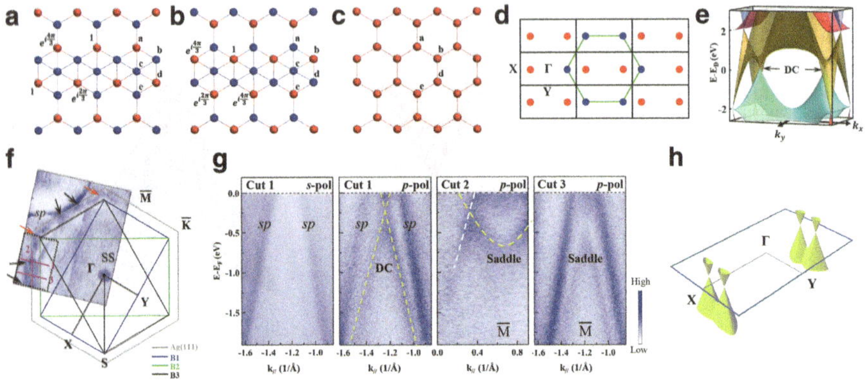

Fig. 4.4 Calculation and ARPES results for β_{12} borophene. (**a**, **b**) Wave functions of each boron atom, as indicated by the amplitudes near the red balls; the blue balls represent the boron atoms with a vanishing amplitude. The boron atoms at site *c* will always have a vanishing amplitude. The red balls are equivalent to the sublattices of the honeycomb lattice shown in (**c**). (**d**) Schematic drawing of the band folding process. The green hexagon and the black rectangles indicate the BZs of the equivalent honeycomb lattice and the β_{12} sheet, respectively. The blue dots represent the original Dirac cones from the honeycomb lattice; the red dots represent the folded Dirac cones. (**e**) Band structures of freestanding β_{12} borophene from the TB model. (**f**) Fermi surface of β_{12} borophene on Ag(111). The black, green, and blue rectangles indicate the BZs of the three equivalent domains; the grey hexagon indicates the BZ of Ag(111). The black and red arrows indicate the bands of the boron layer. The pink lines indicate Cuts 1–3, where the ARPES intensity plots in (**g**) were measured. (**h**) Schematic drawing of the Dirac cones in β_{12} borophene. (Reproduced from Ref. [28])

The discovery of the Dirac cones in β_{12} borophene showed that Dirac cones can exist in non-honeycomb-type 2D lattices. Subsequently, Ezawa et al. derived a Dirac theory for β_{12} borophene [30]. They found that the Dirac cones in β_{12} borophene are gapless, when the inversion symmetry is preserved, while the cones are gapped when the inversion symmetry is broken. Additionally, they found that β_{12} borophene also hosts pseudospin triplet fermions because of the presence of three-band touching points. Lee et al. later studied the bonding characteristics of β_{12} borophene via first-principles calculations and *X*-ray photoelectron spectroscopy (XPS) [31]. They found that the boron atoms at the centers of the honeycombs act as nearly perfect electron donors without forming any additional in-plane bonds, thus forming a hidden honeycomb structure. The novel bonding characteristics of β_{12} borophene resulted in an unusual energy sequence for the core electrons, which has subsequently been verified via XPS measurements [31].

Another borophene phase that has been confirmed to host Dirac cones is the χ_3 phase. Feng et al. used ARPES and first-principles calculations to study the electronic structures of χ_3 borophene and confirmed the existence of a Dirac cone at each corner of the BZ [15]. The fact that these Dirac cones are strongly anisotropic is of particular interest: the highest Fermi velocity is approximately three times that of the lowest Fermi velocity. Additionally, linear dichroism ARPES measurements confirmed that the Dirac bands were mainly derived from the p_z orbitals of boron, similar to the case of β_{12} borophene. These findings indicated that χ_3 borophene can be used to create quantum devices with anisotropic properties.

4.3 Dirac Nodal Lines

In addition to the Dirac cone, the Dirac/Weyl nodal line (Fig. 4.3b) has also been studied intensively in modern condensed matter physics [32, 33]. These unusual topological band structures are protected by crystal symmetries, such as the (glide) mirror reflection symmetry, and are thus resistant to external perturbations. To date, topological nodal lines have been found in numerous 3D materials and have been found to host a variety of exotic properties, including ultrahigh carrier mobility, drumhead surface states, and extremely large magnetoresistance. In the 2D limit, however, such nodal lines are quite rare, and only a few 2D nodal line materials have been found, including monolayer Cu_2Si [34], CuSe [35], and $GdAg_2$ [36]. Therefore, the search for new 2D nodal line materials is also at the forefront of materials science research.

Because of the polymorphism of borophene, the possibility of the existence of Dirac nodal lines has also been studied intensively. Shortly after the discovery of the Dirac cones in β_{12} borophene, Gupta et al. predicted that there was also a Dirac nodal line in β_{12} borophene [37], as shown in Fig. 4.5(a–c). The Dirac nodal line originated from the inversion of the band ordering that occurs along the Γ–X and Γ–Y directions, and it was protected by the mirror reflection symmetry. Interestingly, their calculation results showed that the Dirac nodal line remained intact when the borophene is grown on Ag(111). The coexistence of the nodal lines and the Dirac cones in β_{12} borophene means that its electronic properties are important for both fundamental and applied research.

Another borophene polymorph that hosts Dirac nodal lines is the bilayer P6-mmm borophene, as shown in Fig. 4.5(d–f) [38]. In this phase, the Dirac nodal line is created by the edge states, and this line is also quite robust against both perturbations and impurities. In addition, breaking the crystal symmetries of bilayer P6-mmm borophene could split the nodal line, rather than open a gap. In addition to β_{12} and P6-mmm borophene, Dirac nodal lines have also been predicted to exist in several other 2D boron structures, including P $\bar{6}$-boron [39], hr-sB [40], and β borophene [41]. Despite fruitful theoretical results, experimental observations of such nodal lines in borophenes have not been reported to date.

4.4 Superconductivity

Superconductivity is a phenomenon where a material displays zero electrical resistance and excludes magnetic fields below a critical temperature (T_c). The initial discovery of superconductivity can be dated back to one century ago.

To date, numerous superconductors have been discovered, including conventional Bardeen-Cooper-Schrieffer (BCS)-type superconductors, cuprate superconductors, and iron-based superconductors [42, 43]. In the 2D limit, the superconductivity that accompanies long-range correlation of the order parameter is forbidden by the well-known Mermin-Wagner theory. However, the occurrence of the Kosterlitz-Thouless-Berezinskii (KTB) transition, which is compatible with the Mermin-Wagner theory, allows quasi-long-range correlation of the order parameter to be established and thus makes it possible to realize 2D superconductivity [44, 45]. Recently, superconductivity has been discovered in several 2D systems, including ultrathin metal films, FeSe monolayers on $SrTiO_3$, and magic-angle graphene superlattices [44–46].

The discovery of the high-temperature superconductor MgB_2 ($T_c \approx 39$ K) [47, 48] has stimulated significant research interest in the AB_2 (where $A =$ Mg, Al...) family of materials [49]. Notably, the MgB_2 family of materials consists of alternating honeycomb borophene layers and hexagonal metal layers. The σ-bonding boron orbitals are strongly coupled with the in-plane B-B stretching phonon modes in the borophene layers [50–53], which is essential for the occurrence of high-temperature superconductivity. Therefore, the borophene layers are believed to be the origin of the superconductivity in the MgB_2 family of materials, and this indicates the possible existence of superconductivity in the freestanding borophenes.

No evidence of superconductivity under ambient pressure has been reported for bulk boron, however, despite the existence of a superconducting transition in bulk boron at high pressure. Therefore, the emergence of superconductivity in 2D boron was largely unexpected before the synthesis of the borophenes. Shortly after the experimental realization of the β_{12} and χ_3 borophenes [7, 8], several independent groups reported superconductivity in these borophenes based on theoretical calculations [54–56]. Figure 4.6(a) shows the band structure of freestanding β_{12} borophene [56]. The figure shows that the bonding (σ) and antibonding (σ^*) states are separated by a ~3.5 eV gap and nearly all the σ states are filled. In contrast, the p_z-derived π states are partially filled. As a result, the Fermi surface consists of two $p_{x,y}$-derived hole pockets near Γ, one p_z-derived ellipsoidal electron pocket centered at Y, and a narrow width-modulated strip along the line X–M, as shown in Fig. 4.6(a). The calculated total electron-phonon coupling (EPC) constant is $\lambda \approx 0.8$. The critical temperature can then be estimated using the McMillan equation:

$$k_B T_c = (\hbar\omega_{\ln}/1.2)\exp\left[-1.02\,(1+\lambda)\,/\,\left(\lambda - \mu^* - 0.62\mu^*\lambda\right)\right]$$

where the ω_{ln} prefactor is an appropriately defined logarithmic average frequency [57] and μ^* is a parameter with a typical value of ~0.1 that accounts for the effective Coulomb repulsion. The estimated T_c value is approximately 16 K. Approximately half of λ is caused by coupling to low-frequency phonons with $\omega_{q\nu} \leq 200$ cm^{-1} and particularly to the out-of-plane mode B_{3g}. However, a 20% contribution to λ originates from the higher-frequency optical branches and is mostly related to the A_g and B_{1g} in-plane modes when $\omega > 600$ cm^{-1}. A similar analysis also applies to the χ_3 borophene (see Fig. 4.6(c,d)), and the T_c estimated using the McMillan equation in this case is approximately 12 K.

Apart from the β_{12} and χ_3 borophenes, various borophene polymorphs have been predicted to be superconductors. Zhao et al. reported that conventional BCS superconductivity is a ubiquitous phenomenon in borophene [54, 58] and the T_c values for most borophene polymorphs are above the liquid-helium temperature (~4.2 K). The β_{12} and χ_3 borophenes have estimated T_c values of 13.8 K and 12.1 K, and their calculated electron-phonon coupling constants are 0.82 and 0.67, respectively. These results show qualitative agreement with the results of Penev et al. [56], in spite of slight differences.

Despite the large numbers of theoretical works on the superconductivity of borophenes, no experimental evidence of superconductivity in borophenes has been reported to date. Indeed, the high T_c values were calculated based on freestanding borophenes that are still quite challenging to realize experimentally. To date, the most appropriate substrates for borophene growth are noble metals, such as Ag(111). Taking the β_{12} borophene on Ag(111) as an example, significant charge transfer occurs from the Ag atoms to the B atoms [8, 31]. In addition, the lattice mismatch causes tensile strain in the borophene lattice. These substrate-induced effects, including the charge transfer and strain, are expected to suppress the T_c of borophene [55, 59], thus making it difficult to detect experimentally. However, different values for the calculated T_c of the borophene/Ag(111) structure are given in different papers. For example, Gao et al. reported that the T_c was very high (>10 K) [55], while Cheng et al. reported that the T_c had been suppressed significantly to 0.09 K [59]. Therefore, further theoretical and experimental works are required to clarify this issue. Because borophenes host topological band structures such as the gapless Dirac cones and Dirac nodal lines as discussed in the previous sections, the coexistence of these topological band structures and superconductivity indicates the existence of unusual properties that originate from the interplay between the Cooper pairs and the topological properties in the 2D limit.

Although experimental research into the superconductivity of borophene is still on the way, techniques such as Raman spectroscopy can be used to investigate the vibrational properties and the electron-phonon coupling (EPC) in borophene. The EPC constant in particular is an important parameter for superconductivity. Figure 4.7(a) shows a typical Raman spectrum for the β_{12} borophene on Ag(111) [60]. The spectrum shows multiple Raman peaks, particularly in the low-frequency region. This Raman spectrum is in stark contrast with that of boron powder, which corresponds to the β-rhombohedral phase of bulk boron. When combined with theoretical calculations, the experimental Raman peaks of β_{12} borophene can be assigned and explained well, as shown in Fig. 4.7. For example, the two peaks at 1148 cm^{-1} and 856 cm^{-1} are the $A_g{}^1$ and $A_g{}^2$ modes that originate from in-plane stretching, while the 264 cm^{-1} and 296 cm^{-1} peaks are the $B_{1u}{}^1$ (X) and $B_{1u}{}^2$ modes, respectively (Fig. 4.7).

The EPC constant can be estimated using the Allen formula:

$$\Gamma_i = (1/g_i)\pi N\left(\varepsilon_f\right)\lambda_i \omega_i{}^2$$

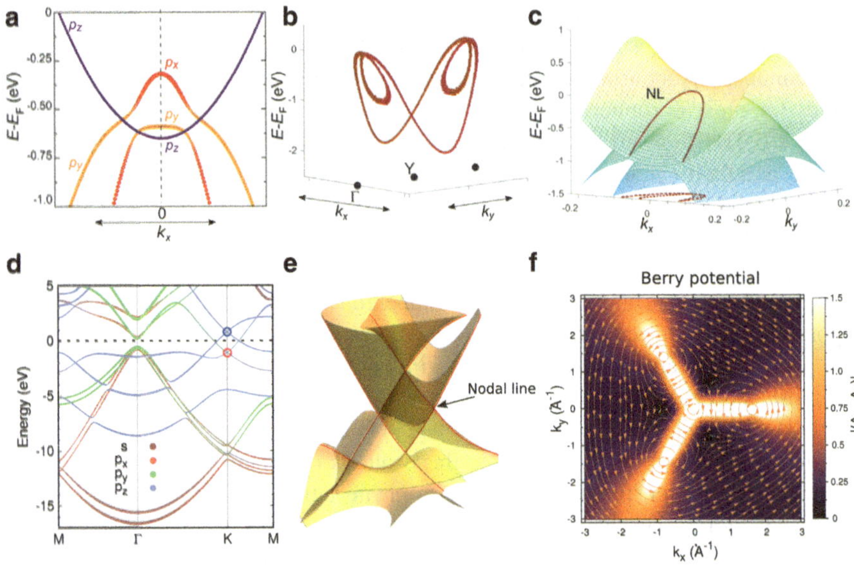

Fig. 4.5 (**a**) Band structure of freestanding β_{12} sheet calculated along k_x (at k_y $(2\pi/b) = 0.15$), showing the inversion in the characteristics of the bands. (**b**) Nodal points extending throughout the 3D energy-momentum space to form a closed loop with a peculiar topology similar to that of a knot. (**c**) 3D band structure showing the band crossing responsible for the nodal line (NL) (marked using a magenta solid line). A projection of the location of the nodal line on the k_x–k_y plane (dotted line) is also shown. (**d**) Band structure of 6B:P6 mmm showing the contributions of the different orbitals using different colors. Blue and red circles represent the Dirac cones that are related to the electrons and holes, respectively. (**e**) 3D plot of the band structure of 6B:P6 mmm calculated by density functional theory near the Fermi level around the K point. (**f**) Berry potential of the valence band for 6B:P6 mmm around the K point. (Reproduced from Ref. [37]. (By courtesy of Boris I. Yakobson). Reproduced from Ref. [38] Reprinted by permission from M. Nakhaee et al. 2018. Copyright (2018) by the American Physical Society)

where Γ_i is the Raman line width, g_i is the degeneracy of the ith mode, ω_i is the unrenormalized discrete-phonon frequency of the ith mode, and $N(\varepsilon_f)$ is the density of states at the Fermi level. The EPC constant calculated from the far-field Raman spectrum is approximately 0.53. This value is comparable to but slightly smaller than the theoretically predicted value for freestanding β_{12} borophene [56]. This slight suppression of the EPC may be caused by the interactions with the Ag(111) substrate. The estimated T_c of the β_{12} borophene based on the McMillan equation is 7.1 K. Similar measurements and analyses of the χ_3 borophene gave an EPC constant of 0.59 and a T_c of 9.8 K.

Tip-enhanced Raman spectroscopy (TERS) is a powerful technique that provides dramatic signal enhancement and high spatial resolution and can be used to detect the vibrational properties of materials on the atomic scale. When the STM tip moves closer to the β_{12} borophene, a dramatically tip-enhanced Raman spectrum can be observed, as shown in Fig. 4.7. The out-of-plane vibrational modes $B_{3g}{}^2$ in the low-

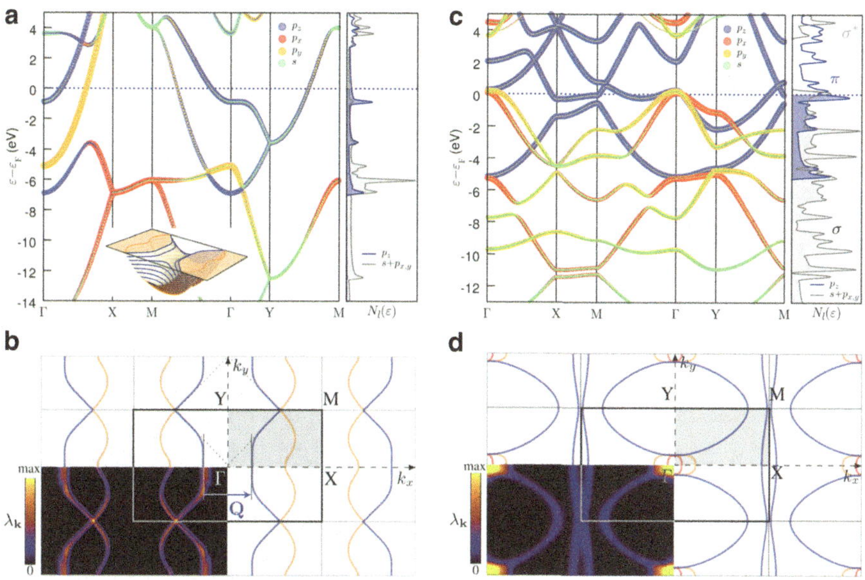

Fig. 4.6 (**a**) Band structure of β_{12} borophene (left panel) and projected density of states (right panel). The orbital characteristics of the bands are thickness and color encoded. (**b**) Fermi surface of the β_{12} borophene in the extended BZ scheme (the first BZ is shown as a bold rectangle with its irreducible part shaded). The separate Fermi contour segments are colored in accordance with their band origins. The lower left quarter shows the magnitude of the effective **k**-resolved electron-phonon coupling parameter λ_k. (**c**) and (**d**) show the same information as (**a**) and (**b**), but for the χ_3 borophene. (Reproduced from Ref. [56]. (By Courtesy of Boris I. Yakobson))

frequency region are strongly enhanced and dominate the spectra. The $B_{1u}{}^1$ and $B_{1u}{}^2$ modes are also significantly enhanced, while the other modes, such as the high-frequency A modes, show negligible enhancements. This selective enhancement effect can be understood well because TERS mainly enhances the out-of-plane vibrational modes [61, 62]. These results confirmed the vibrational modes of β_{12} borophene.

4.5 Edge Magnetism

The edges of graphene can produce additional electronic states near the Fermi level. These states at the Fermi level then lead to the emergence of edge magnetism [63], which has important applications in spintronics. Recently, edge magnetism has also been discovered in several other 2D materials, including silicene, black phosphorus, and transition metal dichalcogenides. Edge magnetism has also been predicted in borophene. Vishkayi et al. studied the electronic and magnetic structures of β_{12} and χ_3 borophene nanoribbons (BNRs) via first-principles calculations [64, 65].

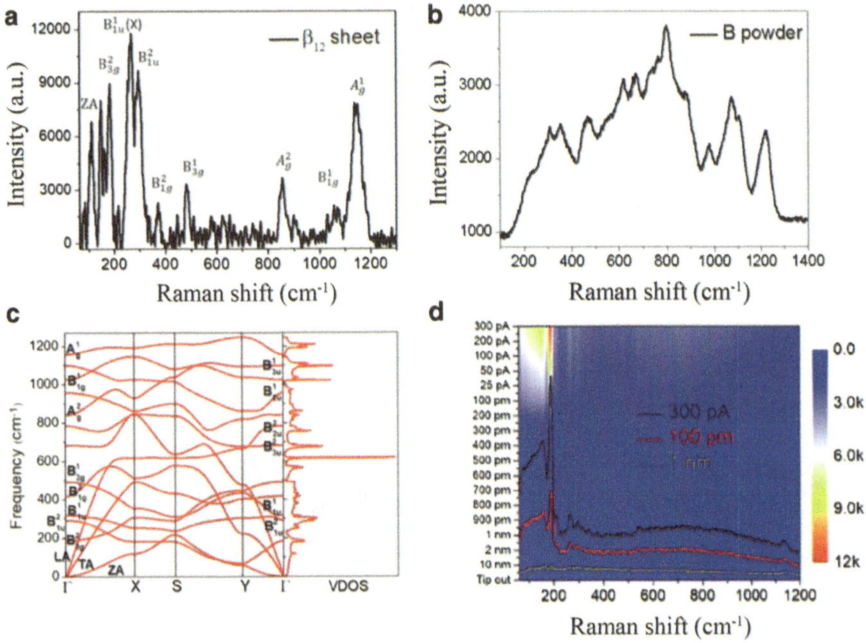

Fig. 4.7 (**a**) In situ far-field Raman spectrum of monolayer β_{12} borophene (laser power of 27.5 mW, data acquisition time of 40 min). (**b**) Raman spectrum of boron powder used to grow the borophene. (**c**) Calculated phonon dispersion curves for β_{12} borophene and the corresponding vibrational density of states. (**d**) Gap distance-dependent TERS spectra of β_{12} borophene on Ag(111) (laser power of 11.5 mW, data acquisition time of 50 s). (Reproduced from Ref. [60])

They found that the magnetic properties of both BNRs were strongly dependent on the cutting direction and the edge profile. Moreover, both BNRs showed strong spin anisotropy at their edges, which indicates their potential for applications in next-generation spintronic devices. Meng et al. studied the electronic and magnetic properties of δ_6 BNRs and found that BNRs with an orientation along one specific direction can adopt a magnetic ground state that is either antiferromagnetic or ferromagnetic, depending on the width of the ribbon [66].

4.6 Chemical Functionalization

Because of its unique bonding structure and chemical reactivity, borophene is prone to adsorption of atoms or molecules that could then change its chemical and physical properties. From this perspective, chemical functionalization provides an effective approach to manipulate the electronic structures of borophenes. For example, hydrogenation can not only improve the stability of metastable borophenes but can also change their electronic structures [67–69]; pristine 2-Pmmn borophene shows a metallic band structure, while fully hydrogenated 2-Pmmn borophene hosts a Dirac cone at its Fermi level along the Γ–X direction in the Brillouin zone [69]. The Fermi velocity is as high as 3.5×10^6 m/s, which is even higher than that of graphene.

In addition, 2D ferromagnetic materials have stimulated considerable interest in materials science research since the recent discovery of intrinsic 2D ferromagnetism in several van der Waals crystals, including $Cr_2Ge_2Te_6$, CrI_3, and VSe_2 [70–72]. Because boron is a nonmagnetic element, there is no intrinsic ferromagnetism in borophenes. However, the functionalization of borophenes using transition metal atoms can produce a long-range ferromagnetic order [73–78]. Although this magnetic order can emerge near the edges or the defects of borophene, almost none of the known borophenes are intrinsically magnetic, which severely hinders their application to spintronic devices. Therefore, the realization of robust ferromagnetism in borophene is highly desirable. Li et al. predicted that the adsorption of 3D transition metal atoms such as Ti, V, Cr, Mn, and Fe on δ_6 borophene can lead to robust ferromagnetism that is dominated by the 3D orbitals of the transition metal atoms [78]. By embedding these transition metal atoms into the intrinsic hexagonal vacancies of borophene, it is possible to realize room-temperature ferromagnetism [77].

In addition, functionalization of borophene using other atoms or molecules has also been explored, and a variety of novel properties have been predicted in these materials [79–83].

4.7 Summary and Outlook

In this chapter, we have seen that borophenes host a variety of unusual properties, including metallicity, topological band structures, and superconductivity. In particular, different novel properties have been shown to coexist within a single borophene polymorph. For example, the β_{12} borophene hosts both gapless Dirac cones and superconductivity, which makes it a promising platform for study of the interplay between Dirac fermions and Cooper pairs in the 2D limit. Therefore, borophenes are of significant importance for both fundamental research and practical applications.

However, there is still a lack of experimental research on borophene, and only a few of the exotic material properties, such as Dirac fermions, have been confirmed experimentally. There are several major difficulties restricting the experimental research into borophene. First, the polymorphism of borophene makes it difficult to synthesize large-scale single-phase borophene specimens. Second, the borophene growth method is limited to molecular beam epitaxy (MBE), and the substrates that can be used are mainly metals. These metallic substrates tend to interact strongly with the borophenes and destroy their intrinsic properties, leading to phenomena such as the suppression of superconductivity in β_{12} and χ_3 borophenes by the substrate. In addition, transport measurements are not feasible in borophene because of the electrical shorting behavior of the metallic substrates. Therefore, the synthesis of large-scale single-phase borophenes on insulating substrates is a prerequisite for transport measurements and the fabrication of borophene-based devices.

References

1. A.R. Organov, V.L. Solozhenko, J. Superhard Mater. **31**, 285 (2009)
2. B. Albert, H. Hillebrecht, Angew. Chem. Int. Ed. **48**, 8640 (2009)
3. Z. Zhang, E.S. Penev, B.I. Yakobson, Chem. Soc. Rev. **46**, 6747 (2017)
4. L. Kong, K. Wu, L. Chen, Front. Phys. **13**, 138105 (2018)
5. D. Li, Y. Chen, J. He, Q. Tang, C. Zhong, G. Ding **27**, 036303 (2018)
6. Z.-Q. Wang, T.-Y. Lu, H.-Q. Wang, Y.P. Feng, J.-C. Zheng, Front. Phys. **14**, 33403 (2019)
7. A.J. Mannix, X.-F. Zhou, B. Kiraly, J.D. Wood, D. Alducin, B.D. Myers, X. Liu, B.L. Fisher, U. Santiago, J.R. Guest, M.J. Yacaman, A. Ponce, A.R. Oganov, M.C. Hersam, N.P. Guisinger, Science **350**, 1513 (2015)
8. B. Feng, J. Zhang, Q. Zhong, W. Li, S. Li, H. Li, P. Cheng, S. Meng, L. Chen, K. Wu, Nat. Chem. **8**, 563 (2016)
9. G.P. Campbell, A.J. Mannix, D.J. Emery, T.-L. Lee, N.P. Guisinger, M.C. Hersam, M.J. Bedzyk, Nano Lett. **18**, 2816 (2018)
10. Q. Zhong, J. Zhang, P. Cheng, B. Feng, W. Li, S. Sheng, H. Li, S. Meng, L. Chen, K. Wu, J. Phys. Condens. Matter **29**, 095002 (2017)
11. W. Li, L. Kong, C. Chen, J. Gou, S. Sheng, W. Zhang, H. Li, L. Chen, P. Cheng, K. Wu, Sci. Bull. **63**, 282 (2018)
12. B. Feng, J. Zhang, R.-Y. Liu, T. Iimori, C. Lian, H. Li, L. Chen, K. Wu, S. Meng, F. Komori, I. Matsuda, Phys. Rev. B **94**, 041408(R) (2016)
13. Y. Liu, E.S. Penev, B.I. Yakobson, Angew. Chem., Int. Ed. **52**, 3156 (2013)
14. Z. Zhang, Y. Yang, G. Guo, B.I. Yakobson, Angew. Commun. **54**, 13022 (2015)
15. B. Feng, J. Zhang, S. Ito, M. Arita, C. Cheng, L. Chen, K. Wu, F. Komori, O. Sugino, K. Miyamoto, T. Okuda, S. Meng, I. Matsuda, Adv. Mater. **30**, 1704025 (2018)
16. J. Zhang, J. Zhang, L. Zhou, C. Cheng, C. Lian, J. Liu, S. Tretiak, J. Lischner, F. Giustino, S. Meng, Angew. Chem. Int. Ed. **57**, 4585 (2018)
17. Z. Liu, M. Zhu, Y. Zheng, J. Mater. Chem. C **7**, 986 (2019)
18. A.H. Castro Neto, F. Guinea, N.M.R. Peres, K.S. Novoselov, A.K. Geim, The electronic properties of graphene. Rev. Mod. Phys. **8**, 109 (2009)
19. M.Z. Hasan, C.L. Kane, Colloquium: Topological insulators. Rev. Mod. Phys. **82**, 3045 (2010)
20. N.P. Armitage, E.J. Mele, A. Vishwanath, Weyl and Dirac semimetals in three- dimensional solids. Rev. Mod. Phys. **90**, 015001 (2018)
21. X.-F. Zhou, X. Dong, A.R. Oganov, Q. Zhu, Y. Tian, H.-T. Wang, Phys. Rev. Lett. **112**, 085502 (2014)
22. A. Lopez-Bezanilla, P.B. Littlewood, Phys. Rev. B **93**, 241405(R) (2016)
23. P. Sengupta, Y. Tan, E. Bellotti, J. Shi, J. Phys. Condens. Matter **30**, 435701 (2018)
24. S.K.F. Islam, J. Phys.: Condens. Matter **30**, 275301 (2018)
25. G.C. Paul, S.K.F. Islam, A. Saha, Phys. Rev. B **99**, 155418 (2019)
26. F. Ma, Y. Jiao, G. Gao, Y. Gu, A. Bilic, Z. Chen, A. Du, Nano Lett. **16**, 3022 (2016)
27. W.-c. Yi, W. Liu, J. Botana, L. Zhao, Z. Liu, J.-y. Liu, M.-s. Miao, J. Phys. Chem. Lett. **8**, 2647 (2017)
28. B. Feng, O. Sugino, R.-Y. Liu, J. Zhang, R. Yukawa, M. Kawamura, T. Iimori, H. Kim, Y. Hasegawa, H. Li, L. Chen, K. Wu, H. Kumigashira, F. Komori, T.-C. Chiang, S. Meng, I. Matsuda, Phys. Rev. Lett. **118**, 096401 (2017)
29. B. Feng, H. Zhou, Y. Feng, H. Liu, S. He, I. Matsuda, L. Chen, E.F. Schwier, K. Shimada, S. Meng, K. Wu, Phys. Rev. Lett. **122**, 196801 (2019)

30. M. Ezawa, Phys. Rev. Lett. **96**, 035425 (2017)
31. C.-C. Lee, B. Feng, M.D. D'angelo, R. Yukawa, R.-Y. Liu, T. Kondo, H. Kumigashira, I. Matsuda, T. Ozaki, Phys. Rev. B **97**, 075430 (2018)
32. C. Fang, H. Weng, X. Dai, Z. Fang, Chin. Phys. B **25**, 117106 (2016)
33. S.-Y. Yang, H. Yang, E. Derunova, S.S.P. Parkin, B. Yan, M.N. Ali, Adv. Phys.-X **3**, 1414613 (2018)
34. B. Feng et al., Nat. Commun. **8**, 1007 (2017)
35. L. Gao et al., Adv. Mater. **30**, 1707055 (2018)
36. B. Feng et al., Phys. Rev. Lett. **123**, 116401 (2019)
37. S. Gupta, A. Kutana, B.I. Yakobson, J. Phys. Chem. Lett. **9**, 2757 (2018)
38. M. Nakhaee, S.A. Ketabi, F.M. Peeters, Phys. Rev. B **98**, 115413 (2018)
39. S.-G. Xu, B. Zheng, H. Xu, X.-B. Yang, J. Phys, Chem. C **123**, 4977–4983 (2019)
40. H. Zhang, Y. Xie, Z. Zhang, C. Zhong, Y. Li, Z. Chen, Y. Chen, J. Phys. Chem. Lett. **8**, 1707 (2017)
41. X. Zuo et al., Phys. Rev. B **100**, 115423 (2019)
42. B.T. Matthias, T.H. Geballe, V.B. Compton, Rev. Mod. Phys. **35**, 1 (1963)
43. B. Keimer, S.A. Kivelson, M.R. Norman, S. Uchida, J. Zaanen, Nature **518**, 179 (2015)
44. Y. Saito, T. Nojima, Y. Iwasa, Nat. Rev. Mater. **2**, 16094 (2016)
45. T. Uchihashi, Supercond. Sci. and Technol. **30**, 013002 (2016)
46. Y. Cao, V. Fatemi, S. Fang, K. Watanabe, T. Taniguchi, E. Kaxiras, P. Jarillo-Herrero, Nature **556**, 43 (2018)
47. J. Nagamatsu, N. Nakagawa, T. Muranaka, Y. Zenitani, J. Akimitsu, Nature **410**, 63 (2001)
48. W.N. Kang, H.-J. Kim, E.-M. Choi, C.U. Jung, S.-I. Lee, Science **292**, 1521 (2001)
49. M. Mudgel, V.P.S. Awana, H. Kishan, I. Felner, G.A. Alvarez, G.L. Bhalla, J. Appl. Lett **105**, 07E313 (2009)
50. K.-P. Bohnen, R. Heid, B. Renker, Phys. Rev. Lett. **86**, 5771 (2001)
51. T. Yildirim, O. Gülseren, J.W. Lynn, C.M. Brown, T.J. Udovic, Q. Huang, N. Rogado, K.A. Regan, M.A. Hayward, J.S. Slusky, T. He, M.K. Haas, P. Khalifah, K. Inumaru, R.J. Cava, Phys. Rev. Lett **87**, 037001 (2001)
52. Y. Kong, O.V. Dolgov, O. Jepsen, O.K. Andersen, Phys. Rev. B **64**, 020501(R) (2001)
53. H.J. Choi, D. Roundy, H. Sun, M.L. Cohen, S.G. Louie, Nature **418**, 758 (2002)
54. Y. Zhao, S. Zeng, J. Ni, Appl. Phys. Lett. **108**, 242601 (2016)
55. M. Gao, Q.-Z. Li, X.-W. Yan, J. Wang, Phys. Rev. B **95**, 024505 (2017)
56. E.S. Penev, A. Kutana, B.I. Yakobson, Nano Lett. **16**, 2522 (2016)
57. P.B. Allen, R.C. Dynes, Phys. Rev. B **12**, 905 (1975)
58. Y. Zhao, S. Zeng, J. Ni, Phys. Rev. B **93**, 014502 (2016)
59. C. Cheng, J.-T. Sun, H. Liu, H.-X. Fu, J. Zhang, X.-R. Chen, S. Meng, 2D Mater **4**, 025032 (2017)
60. S. Sheng, J.-B. Wu, X. Cong, Q. Zhong, W. Li, W. Hu, J. Gou, P. Cheng, P.-H. Tan, L. Chen, K. Wu, ACS Nano **13**, 4133 (2019)
61. R. Zhang et al., Nature **498**, 82 (2013)
62. S. Sheng, J.-B. Wu, X. Cong, W. Li, J. Gou, Q. Zhong, P. Cheng, P.-H. Tan, L. Chen, K. Wu, Phys. Rev. Lett. **119**, 196803 (2017)
63. M. Fujita, K. Wakabayashi, K. Nakada, K. Kusakabe, J. Phys. Soc. Jpn. **65**, 1920 (1996)
64. S.I. Vishkayi, M.B. Tagani, Nano-Micro Lett. **10**, 14 (2018)
65. S.I. Vishkayi, M.B. Tagani, Phys. Chem. Chem. Phys. **20**, 10493 (2018)
66. F. Meng, X. Chen, S. Sun, J. He, Phys. E. **91**, 106 (2017)
67. Y. Jiao, F. Ma, J. Bell, A. Bilic, A. Du, Angew. Chem. Int. Ed. **55**(35), 10292 (2016)
68. Z. Wang, T.Y. Lu, H.Q. Wang, Y.P. Feng, J.C. Zheng, Phys. Chem. Chem. Phys. **18**(46), 31424 (2016)
69. L.C. Xu, A. Du, L. Kou, Phys. Chem. Chem. Phys. **18**(39), 27284 (2016)
70. C. Gong et al., Nature (London) **546**, 265 (2017)
71. B. Huang et al., Nature (London) **546**, 270 (2017)

72. M. Bonilla, S. Kolekar, Y. Ma, H.C. Diaz, V. Kalappattil, R. Das, T. Eggers, H.R. Gutierrez, M.-H. Phan, M. Batzill, Nat. Nanotechnol. **13**, 289 (2018)
73. J.C. Alvarez-Quiceno, G.R. Schleder, E. Marinho Jr., A. Fazzio, J. Phys. Condens. Matter **29**, 305302 (2017)
74. X. Zhang, Y. Sun, L. Ma, X. Zhao, X. Yao, Nanotechnology **29**, 305706 (2018)
75. S. Tomar, P. Rastogi, B.S. Bhadoria, S. Bhowmick, Y.S. Chauhan, A. Agarwal Eur, Phys. J. B **91**, 51 (2018)
76. J. Jiang, X. Wang, Y. Song, Comput. Mater. Sci. **153**, 10 (2018)
77. X. Li, X. Li, J. Yang, J. Phys. Chem. Lett. **10**, 4417 (2019)
78. J.Y. Li, H.Y. Lv, W.J. Lu, D.F. Shao, R.C. Xiao, Y.P. Sun, Phys. Lett. A **380**, 3928 (2016)
79. A.A. Kistanov, Y. Cai, K. Zhou, N. Srikanth, S.V. Dmitriev, Y.-W. Yang, Nanoscale **10**, 1403 (2018)
80. M. Shahrokhi, Comput. Mater. Sci. **156**, 56 (2019)
81. T. Liu, Y. Chen, M. Zhang, L. Yuan, C. Zhang, J. Wang, J. Fan, AIP Adv. **7**, 125007 (2017)
82. C.-S. Huang, A. Murat, V. Babar, E. Montes, U. Schwingenschlogl, J. Phys. Chem. C **122**, 14665 (2018)
83. R. Pekoz, M. Konuk, M.E. Kilic, E. Durgun, ACS Omega **3**, 1815 (2018)

Chapter 5
Chemically Modified Borophene

Takahiro Kondo and Iwao Matsuda

5.1 Hydrogenated Borophene (Borophane)

Among two-dimensional (2D) materials that have great potential in various applications [1–3], boron-related materials have unique characteristics of exhibiting polymorphisms [4–6] owing to the ability to form multi-center bonding configurations of boron [7]. There are a wide variety of stable 2D phases of borophene [7–12] and chemically modified borophene. These layers are expected to have different functionalities and to be building blocks from the viewpoint of large-scale material production [13]. Here, we start our chapter with a description of hydrogenated borophene (borophane) [6].

In previous studies, the 2D polymorph phases of borophane [6, 14–16], hydrogenated boron clusters [17], and boron monohydride under high pressure [18] have been theoretically predicted. Borophane with a local structure corresponding to a predicted structure has been then experimentally synthesized in powder form [19–22]. Herein, theoretically predicted borophane sheets, experimentally synthesized borophane sheets, and the predicted applications of borophane are described.

5.1.1 Theoretically Predicted Borophane Sheets

In 2011, the Zhang group proposed a borophane structure as a layered hydrogenated borophene using first-principles calculations [14, 15]. The geometric and

T. Kondo (✉)
Faculty of Pure and Applied Sciences, University of Tsukuba, Tsukuba, Ibaraki, Japan
e-mail: takahiro@ims.tsukuba.ac.jp

I. Matsuda
The Institute for Solid State Physics, The University of Tokyo, Kashiwa, Chiba, Japan
e-mail: imatsuda@issp.u-tokyo.ac.jp

I. Matsuda, K. Wu (eds.), *2D Boron: Boraphene, Borophene, Boronene*,
https://doi.org/10.1007/978-3-030-49999-0_5

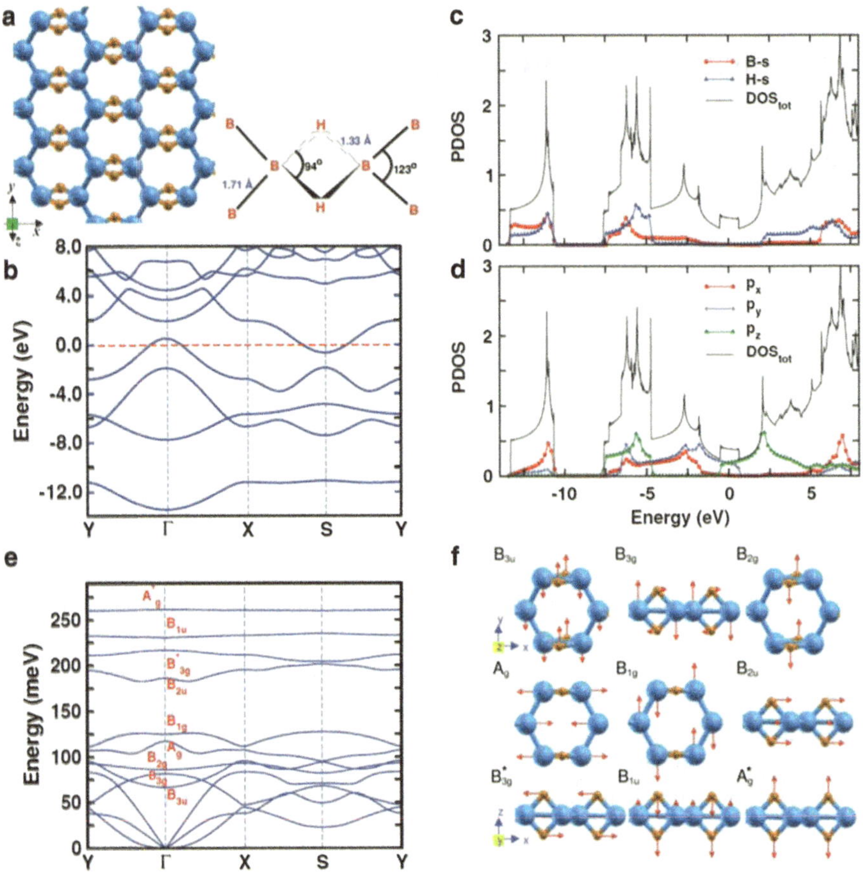

Fig. 5.1 Theoretical prediction of single-layer B_2H_2 [14, 15]. (**a**) The relaxed structure with bond angles and bond lengths. (**b**) The band structure. The Fermi level is at 0 eV and is shown with the dotted line. (**c**) Projected density of states (PDOS) for B 2*s* and H 1*s* and (**d**) that for B $2p_x$, B $2p_y$ and B $2p_z$. The Fermi level is at 0 eV. Units are in states/eV/cell. Total density of states (DOS_{tot}) is shown together. (**e**) The phonon dispersion. The labels at Γ are the nine optical phonon modes. (**f**) The phonon polarization vectors of the optical phonons at Γ. Boron is denoted by large spheres and hydrogen by small spheres. (Reprinted by permission from Abtew et al. [14]. Copyright (2011) by the American Physical Society. Reprinted by permission from Abtew and [15]. Copyright (2011) by the American Physical Society)

band structures of the borophane, single-layer B_2H_2, are shown in Fig. 5.1a, b, respectively. The structure consists of a graphene-like hexagonal boron network and bridge hydrogens, which form 3c2e-like multicenter B–H bonds. Compared to the band structure of MgB_2, which consists of Mg cations and an anionic graphene-like hexagonal boron network, that of borophane shows that it is metallic as MgB_2, but the degeneracy of the p-σ bands at the Γ point in MgB_2 is reduced following the lowered symmetry from D_{6h} to D_{2h}. As a result, only one p-σ band crosses the Fermi level and the hole pocket near the Γ point comes from the upper p-σ band. The degeneracy of the p-π bonding and antibonding states near the Fermi level is also removed. Similar to MgB_2, the Fermi surface of B_2H_2 consists of both a p-σ derived hole pocket and a p-π derived electron pocket. The projected density of states (PDOS) also exhibits significant involvement of the hydrogen σ states in bonding (Fig. 5.1c). The electronic states near the Fermi level are derived from bonding B-p_y and antibonding B-p_z states (Fig. 5.1d). Further detail of the semimetal nature of the borophane will be described in Section §5.1.3. The phonon dispersion of single-layer B_2H_2 does not exhibit any soft phonon modes (i.e., imaginary phonon energies), thus confirming the dynamical stability of the proposed structure (Fig. 5.1e, f). The Zhang group has also predicted that charging the structure significantly softens hydrogen-related phonon modes ($A_g{}^*$, B_{1u}, $B_{3g}{}^*$, and B_{3g} phonons). Boron modes, in contrast, are either hardened or not significantly affected by electron doping. Furthermore, self-doping the structure (bonding with additional boron atoms) considerably reduces the energy barrier against hydrogen release. The results suggest that electrochemical charging or self-doping mechanisms may facilitate hydrogen release while maintaining the underlying boron network for subsequent rehydrogenation.

In 2016, the Du group reported other stable phases of borophane [6]. Figure 5.2 shows the predicted structures. In every case, no imaginary phonon mode can be found in the first Brillouin zone, indicating their dynamical stability. The "ladder-like cluster [17] extended BH layers" (*C2/m*, Fig. 5.2a) have 3c2e-like multicenter B–H–B bonds like B_2H_2 in Fig. 5.1, while single B–H bonds are apparently formed in a multicenter bonding configuration of boron for the "buckled B_7 cluster [23] extended BH sheet" (*Pbcm*, Fig. 5.2b) and "hydrogenated borophene (triangular monolayer borophene) [8]" (*Pmmn*, Fig. 5.2c). All the structures shown here (*C2/m*, *Pbcm*, and *Pmmn*) have been predicted to have a characteristic Dirac cone in their electronic band structure, indicating an intriguing structure for device applications. In addition, their mechanical properties have also been examined by estimating the 2D Young's modulus and Poisson's ratio, revealing the high mechanical strength of these borophane sheets.

The *Pmmn* borophane (Fig. 5.2c) structure has also been reported [24–27]. Kou et al. reported that borophane with the structure shown in Fig. 5.3a–c has direction-dependent Dirac cones [24]. Specifically, as shown in Fig. 5.3d, most states near the Fermi level were removed after hydrogenation, and the remaining band states exhibit perfect linear dispersion along Γ to X, with the valence and conduction bands meeting at a single point at the Fermi level. The Dirac fermions possess an ultrahigh Fermi velocity of up to $3.5 \times 10^6\ \mathrm{ms}^{-1}$, which is 4 times higher than that of graphene. The Dirac cone is distorted and asymmetric as shown in Fig. 5.3e. The two crossing bands at the Fermi level are mainly caused by the p_x and p_y orbitals of the boron atoms (Fig. 5.3f, g). In the real space, the Bloch states at the Dirac points clearly confirm that in-plane p_x and p_y orbitals lead to the formation of a Dirac cone. The bonds between p_x orbitals form a σ-conjugated linear chain, while the coupling between p_y orbitals leads to the formation of a π bonding-like framework. The energy band crossings between p_x and p_y are responsible for the production of the Dirac cones. In particular, the Dirac cones of most 2D hexagonal materials mainly

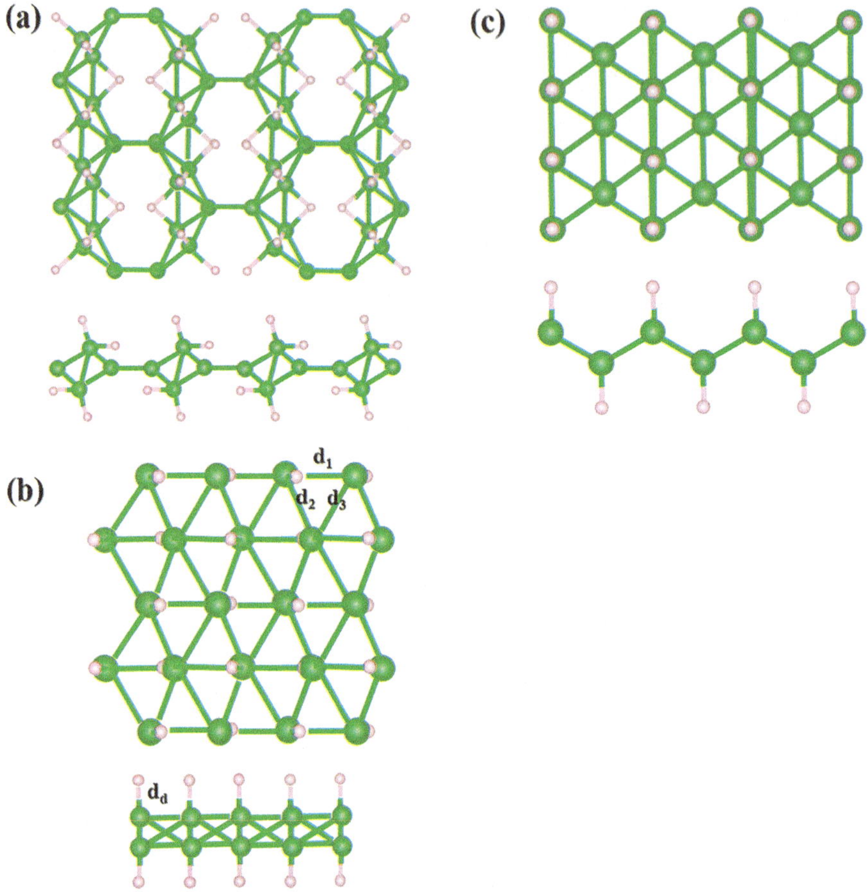

Fig. 5.2 (**a–c**) *C2/m*, *Pbcm*, and *Pmmn* structures, top and side views. (Drown based on Ref. [6])

originate from the orbitals along the vertical direction of the 2D plane, such as the p_z orbital of graphene. In borophane, however, its Dirac cone is formed from in-plane orbitals but not orbitals along the z direction. Kou et al. concluded that in-plane state-induced Dirac states are expected to be stable and tunable under in-plane strain but robust against vertical electric fields. In their study, Young's moduli were 190 and 120 GPa nm along the x (armchair) and y (zigzag) directions in Fig. 5.3a, which are comparable to those of steel. The ultrahigh Fermi velocity and good mechanical features thus indicate that borophane is ideal for nanoelectronic applications [24]. Padilha et al. reported the transport property of borophane with the same structure, which was evaluated through Landauer–Büttiker formalism using a first-principles calculation study. They found that the electronic current is two orders of magnitude smaller for borophane compared to borophene [25]. The electronic current was found to be approximately double along the x (armchair) direction compared to

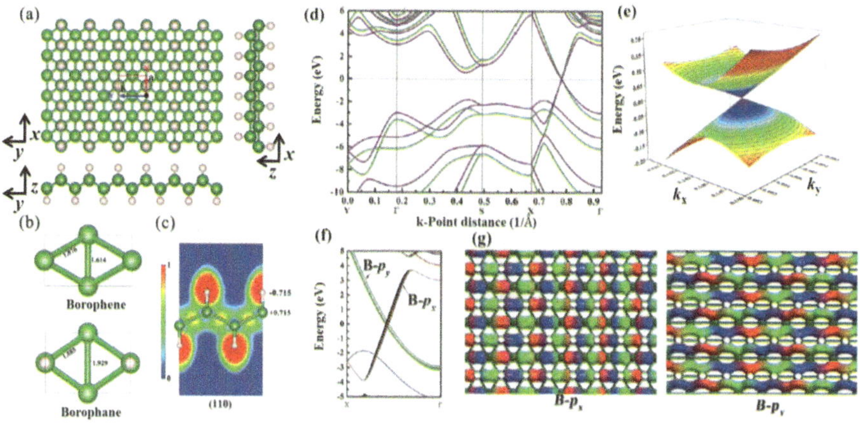

Fig. 5.3 Geometric and electronic structure of borophane [24]. (**a**) Top and side views of the optimized borophane. The unit cell is marked with a red box. The large (green) and small (white) balls represent B and H atoms, respectively. (**b**) Typical B–B bond lengths in borophane and borophene unit cells where the unit of bond length is Å. (**c**) ELF (electron localization function) schematic of the borophane (110) surface where the electron transfer is shown after hydrogenation in the charge unit of e. (**d**) The band structures of borophane based on the Perdew–Burke–Ernzerhof (PBE, green curves in figure) and Heyd–Scuseria–Ernzerhof (HSE06, purple curves in figure) hybrid functionals where the Fermi level is shifted to zero. (**e**) Distorted 3D Dirac cone formed in the vicinity of the Dirac point. (**f**) Orbitals of the linear dispersion relation along Γ to X. The Fermi level is set to zero. (**g**) Bloch states of the p_x and p_y orbitals of boron atoms at the Dirac points of borophane. (Reprinted by permission from Xu et al. [24]. Copyright (2016) ROYAL SOCIETY OF CHEMISTRY)

that along the y (zigzag) direction in Fig. 5.3a. The stretching of the borophane sheet was suggested to increase this direction dependence of the electronic current of borophane [25]. Zheng et al. reported that the mechanical properties and phonon stability of this borophane are both highly anisotropic [27]. Specifically, the ultimate tensile strain along the x (armchair) direction is only 0.12, but it can be as large as 0.3 along the y (zigzag) direction in Fig. 5.3a. Wang et al. reported that the intrinsic lattice thermal conductivity of this borophane is also highly anisotropic and is twice as high along the y (zigzag) direction compared to that along the x (armchair) direction in Fig. 5.3a [28]. In 2018, the tight-binding model for this borophane in the two-center approximation was reported, where the anisotropic Dirac cone described above is found in the calculated band structure of borophane [29].

In 2017, Zheng et al. reported four types of new stable hydrogenated borophane conformers (B-borophane, TCB-borophane, T-borophane, and W-borophane) as shown in Fig. 5.4, based on first-principles calculations [16]. Compared with chair-like borophane (C-boropane, the same structure as Figs. 5.2c and 5.3a), all four new conformers were found to be much lower in total energy. In the calculated phonon dispersions, no imaginary frequencies along the high-symmetry directions of the Brillouin zone were found, indicating that they are all dynamically stable. By analyzing the atomic arrangements and the total energies in different conformers,

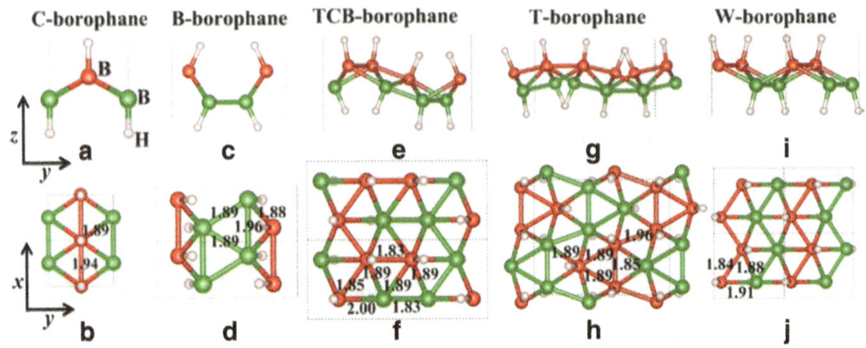

Fig. 5.4 Crystal structures of borophane [16]. Side and top views of: (**a**, **b**) C-borophane, (**c**, **d**) B-borophane, (**e**, **f**) TCB-borophane, (**g**, **h**) T-borophane, and (**i**, **j**) W-borophane. The numbers are the B–B bond lengths in Å. The large red and green balls represent B atoms; the small light pink balls represent H atoms. (CC BY: Wang et al. [16])

they found that the configuration in which B atoms are staggered in the zigzag mode along the *y* direction and in the up-and-down wrinkle mode along the *x* direction is the most stable (Fig. 5.4). The most stable conformer of W-borophane has an energy difference of about 113.41 meV/atom lower than C-borophane. The band structures of the four new conformers all have a Dirac cone along the Γ-Y or Γ-X direction.

5.1.2 Experimental Synthesis of Borophane Sheets

In 2017, Kondo et al. reported the experimental synthesis of borophane by ion-exchange between protons and Mg cations in MgB_2 [19]. Specifically, as shown in Fig. 5.5a, ion-exchange was conducted by adding MgB_2 powder to methanol (or acetonitrile) in the presence of an ion-exchange resin; this entire process was conducted with nitrogen at room temperature and ambient pressure with stirring because MgB_2 is sensitive to the presence of water [30]. Hydrogen boride (HB) sheets were then obtained in the form of a yellow powder as the ion-exchange reaction product at an average reproducible yield of 42.3%.

Scanning electron microscopy (SEM) images of the HB sheets exhibit clear bent and/or folded sections as shown in Fig. 5.5b, indicating the flexible nature of the exfoliated 2D sheets. No Mg signal was detected from the HB sheets by X-ray photoelectron spectroscopy (XPS), as shown in Fig. 5.5c, while a significant amount of hydrogen (H_2) was detected as the major desorption species in thermal desorption spectroscopy (TDS) measurements (Fig. 5.5d). These results indicate that the Mg cations of MgB_2 were completely exchanged for protons. Considering the TDS results and weights for 10 different sample lots, the elemental ratio of the HB sheets was determined to be H:B = 1:1.02 ± 0.18. These results show that HB sheets with a H:B ratio of 1 were formed by following a stoichiometric ion-exchange reaction at room temperature:

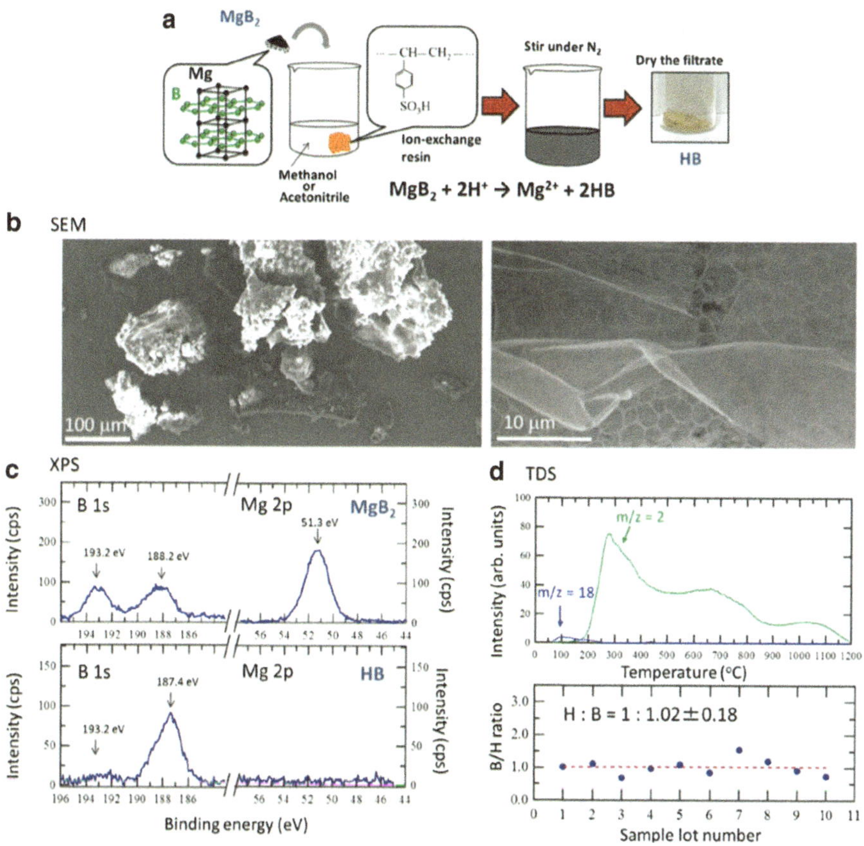

Fig. 5.5 Ion-exchange between protons and Mg cations of MgB_2 produces HB sheets [19]. (**a**) Synthesis scheme for HB sheets. (**b**) SEM images of HB sheets. Left: powders of HB sheets mounted on Si wafer. Right: HB sheets obtained by drying the supernatant of the solution shown in panel (**a**) on a TEM grid. (**c**) XPS B 1s and Mg 2p core-level spectra for MgB_2 and HB sheets. (**d**) TDS of HB sheets, where the elemental ratio of H:B is approximately 1:1 based on TDS data and weights for 10 different sample lots

$$MgB_2 + 2H^+ \rightarrow Mg^{2+} + 2HB$$

It is well-known that XPS core-level binding energies reflect the charge states of surface elements in terms of chemical shift and from the XPS B 1s spectra the charge states of surface boron atoms before and after the ion-exchange process were determined. As shown in Fig. 5.5c, there are two peaks in the B 1s region for the starting material (MgB_2). The peak at the lower binding energy (188.2 eV) corresponds to the negatively charged boron species in MgB_2, while the peak at the higher binding energy (193.2 eV) corresponds to the positively charged boron-forming boron oxides such as B_2O_3 on the MgB_2 surface, as reported previously [31]. The Mg 2p peak from MgB_2 at 51.3 eV is also in accordance with the value reported in the literature, which is known to reflect the positive Mg charge of greater than 0 but less than +2 (B is thus negatively charged) [31–33]. Note that the negatively charged boron species can be clearly observed after the ion-exchange process (HB sheets) as a peak at 187.4 eV. The presence of a negatively charged boron peak without the Mg peak in Fig. 5.5c further indicates that the Mg cations were exchanged with protons by this ion-exchange process to form HB sheets.

The sheets feature an sp^2-bonded planar boron structure without any long-range order. A hexagonal boron network with bridging hydrogens (structure shown in Fig. 5.1) is then suggested as the possible local structure, where the absence of long range order was ascribed to the presence of three different anisotropic domains originating from the twofold symmetry of the hydrogen positions against the sixfold symmetry of the boron networks, based on X-ray diffraction, X-ray atomic pair distribution functions, electron diffraction, transmission electron microscopy, photo absorption, core-level binding energy data, infrared absorption, electron energy loss spectroscopy, and density functional theory calculations [19]. Quite recently, an improved structure analysis for pure HB sheets was reported together with atomic force microscopy analysis as well as conductivity measurements [21].

It is interesting to focus on the charge state of boron and hydrogen in this particular borophane (graphene-like borophane or HB sheets) because it is counterintuitive from the electronegativity difference between boron and hydrogen. In the case of experimentally synthesized HB sheets (Fig. 5.5), the charge state of boron is negative because the XPS B 1s peak is located at 187.4 eV, which is almost the same position as (or a slightly more negatively charged position compared to) that of MgB_2 (188.2 eV) (Fig. 5.5c). Moreover, the HB sheets are prepared by ion-exchange between protons and magnesium cations. Thus, the HB sheets shown in Fig. 5.5 must be called "hydrogen boride" rather than the traditional borane name of "boron hydride." The charge states represented by the B 1s core level have also been shown using first-principles calculations. Specifically, as shown in Fig. 5.6, only the derived local structure (and the structure named borophane-I) shows a deep position of the B 1s state compared to that of the other borophane structures, which is consistent with the experimental result that it is composed of negatively charged boron (herein, the discussion was based on the B 1s core-level states rather than the charge estimated using the theoretically calculated Bader charge [34]). On the other hand, the structure shown as borophane-III corresponds to the structures shown in Fig. 5.4a (C-borophane), Figs. 5.2c and 5.3a, where clear charge transfer from boron to hydrogen was reported [16, 24, 35] (e.g., see the charge states in Fig. 5.3c), which qualitatively agrees with the shallow B1s energy in Fig. 5.6. By comparing the electronegativity of boron and hydrogen, we can see that the electronegativity of hydrogen (2.20) is larger than that of boron (2.04). Therefore, the charge transfer from B atoms to H atoms is reasonable for borophane-III. Similarly, charge transfer from silicon (1.90) to hydrogen (2.20) atoms has been found in hydrogenated silicene [36]. However, although the electronegativity of carbon (2.55) is larger than that of hydrogen (2.20), charge transfer from H atoms to C atoms has been observed in graphane [37]. Thus, we should not determine charge state only from the electronegativity difference.

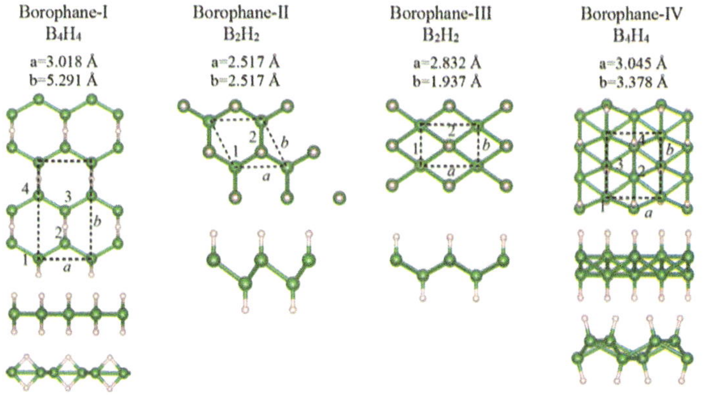

	Sample	B 1s (eV)	Relative energy difference (eV)
Experiment	$B(OH)_3$	193.2	0
	Hydrogen boride sheets	187.4	-5.8
DFT calculation	$B(OH)_3$	202.4	0
	Local HB structure	196.2	-6.2
	Borophane-I (B_4H_4, *Cmmm*)	197.3	-5.1
	Borophane-II (B_2H_2, *P-3m1*)	198.7	-3.7
	Borophane-III (B_2H_2, *Pmmn*)	199.1	-3.3
	Borophane-IV (B_4H_4, *CPbcm*)	199.4	-3.0
	Borophene	201.2	-1.2

Fig. 5.6 B 1s core-level binding energies of HB Sheets [19]. The optimized structure used for the calculation is shown above. The relative energy difference compared with B 1s of $B(OH)_3$ is shown, since the absolute value needs careful treatment such as correction of the so-called final state effect (effect of incomplete screening of created holes during photoemission)

In the case of the hydrogen boride sheet, the boron energy position is deeper compared to that of a single boron atom due to the formation of a huge boron network with sp^2-like bonding configurations (boron–boron hybridizations) [38] and the boron atoms bond with hydrogen atoms via three-center two-electron bonds, which are composed by occupied hydrogen orbital and occupied and unoccupied orbitals of boron. Consequently, the boron network is negatively charged. Detail of this picture will be carefully examined in our future work. At least, the experimentally measured acid function of the HB sheet (below 1.5 and above 0.43) [19] and the experimentally revealed solid-acid catalytic property of the HB sheets [39] indicate the protonic state of hydrogen in the HB sheets.

5.1.3 Semimetallicity of Borophane Sheets

During the formation of graphene-like borophane (HB) sheets (structure shown in Fig. 5.1), hydrogenation of the borophene layers energetically results in the three-

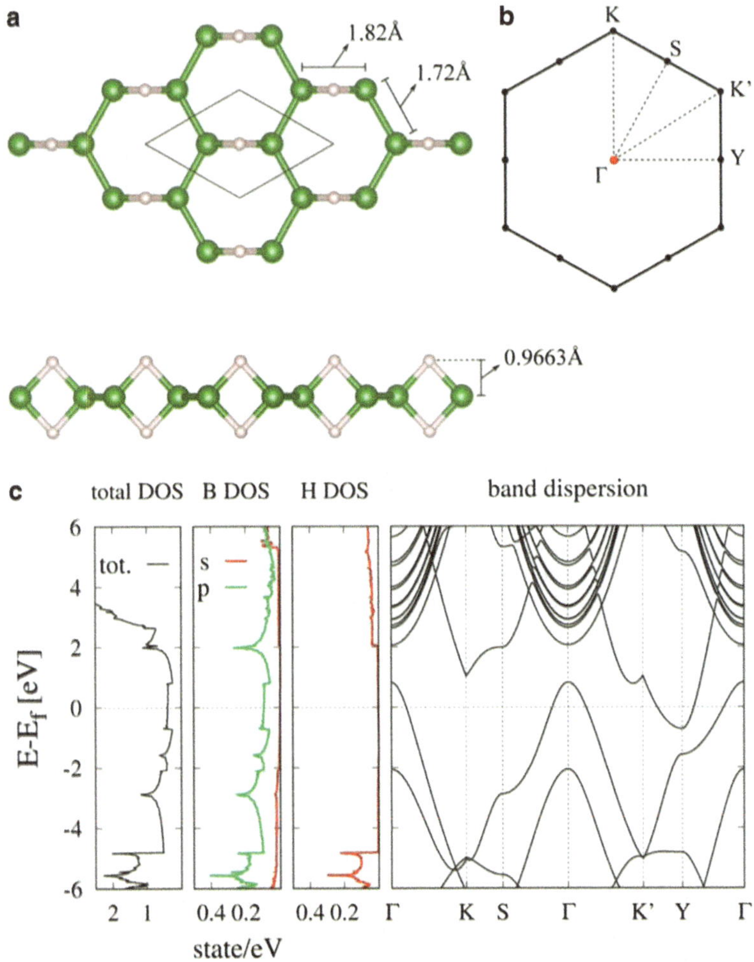

Fig. 5.7 (**a**) A structure model of the HB layer (borophane). Boron and hydrogen atoms are indicated by green and light pink balls, respectively. Drawings of chemical bonds are defined by interatomic distance. (**b**) 2D Brillouin zone and (**c**) calculated band diagram between the symmetry points in the Brillouin zone. (Reprinted by permission from Tateishi et al. [38]. Copyright (2019) by the American Physical Society)

center two-electron bond between the H and B atoms, as realized in a diborane (B_2H_6) molecule. In this case, the HB structure is built with the B_2H_2 blocks with even numbers of valence electrons, $2(N\text{B} + N\text{H}) = 12$, where NB and NH are the number of valence electrons in a boron atom ($N\text{B} = 5$) and a hydrogen atom ($N\text{H} = 1$), respectively. Thus, a HB sheet has an even number of electrons in a unit cell and the electronic structure is semimetal or insulating.

Figure 5.7 gives an example of the band diagram of a HB sheet that is composed of six-membered rings. There is a hole pocket and an electron pocket at the Γ

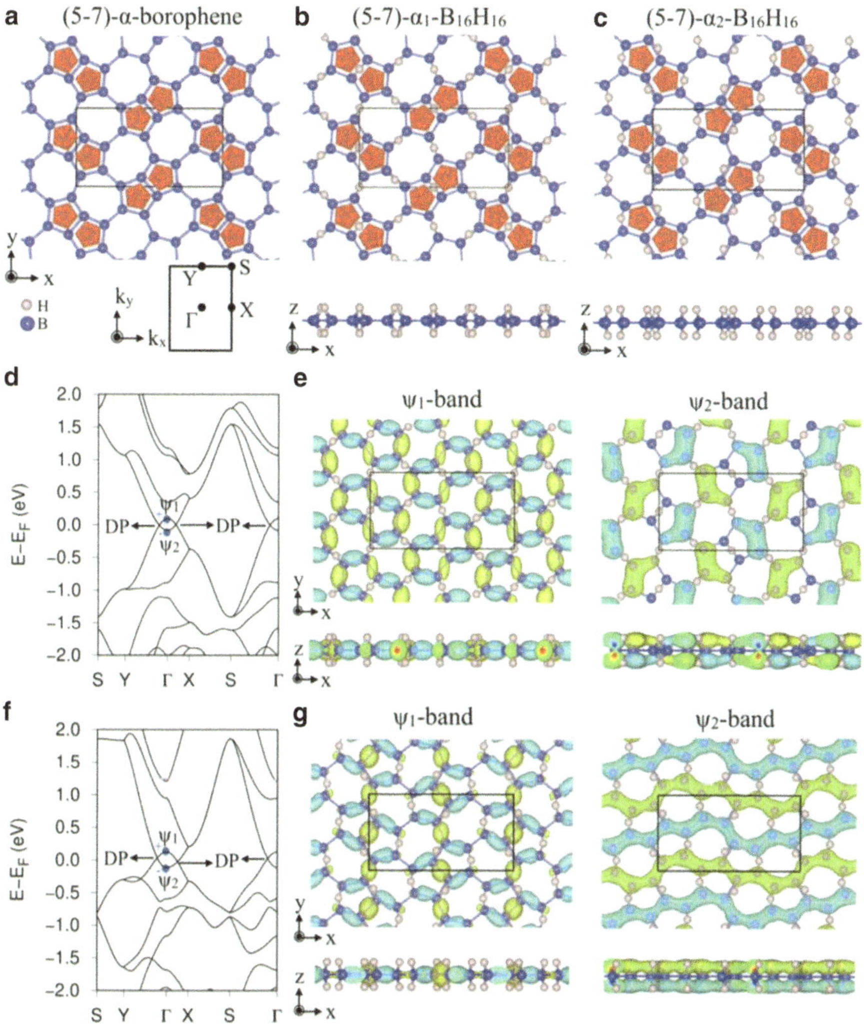

Fig. 5.8 Structure models of monolayers of (**a**) (5-7)-α-borophene, (**b**) (5-7)-α_1-$B_{16}H_{16}$, and (**c**) (5-7)-α_2-$B_{16}H_{16}$. Boron and hydrogen atoms are shown as blue and light pink balls, respectively. The 2D Brillouin zone and the unit cell are shown in the figure. Locations of the five-membered rings are indicated with red pentagons [78]. (**d, f**) Electronic band diagrams and (**e, g**) spatial distributions of wave functions of Ψ_1 and Ψ_2 bands at the Γ point, respectively. Dirac points of the nodal loop are indicated as DP. The parity of mirror reflection symmetry with respect to the x–y plane (layer plane) for the Ψ_1 and Ψ_2 bands are labeled with + and − signs, respectively. The sign of wave functions is indicated by the light blue and light green color [78]

and Y points, respectively, confirming the semimetallic nature [14, 38]. Figure 5.8 shows other examples. Two types of borophane are formed by hydrogenation of a borophene of the five-membered and seven-membered rings, and they have different configuration of hydrogen atoms. We name them (5-7)-α_1-$B_{16}H_{16}$ and (5-7)- α_2-$B_{16}H_{16}$. The 2D band structures, presented in Fig. 5.8, again show the semimetallicity of borophane. It is worth noting that dispersion curves of the hole- and electron-pockets cross each other and form the Dirac points (DP). In the 2D Brillouin zone, the DP are arranged in a circle around the Γ point, and forms the Dirac nodal loop (DNL). Existence of the DNL can be understood from wave functions of the bands, as depicted in Fig. 5.8. Since borophane maintains mirror symmetry with respect to the layer plane, there are even and odd parities for the electronic states. A wave function of the Ψ_1 band has even symmetry, while that of the Ψ_2 band has odd symmetry. The difference in the parity symmetrically restricts hybridization between the two types of electronic states, making the crossing dispersion curves.

An argument with respect to symmetry develops further to intriguing predictions of DNL, Fig. 5.9a, based on topology. The discussion requires a little knowledge of the space groups. When the structure of a crystal belongs to a non-symmorphic space group, P_{bam} (layer group No. 44 or 3D space group No. 55), the electronic states have even and odd parities caused by mirror symmetry with respect to the x–y plane, as in the case of Fig. 5.8. When the electronic bands are doubly degenerate, it has also been proven that the two degenerate bands have the same parity. For example, along the zone boundaries, including the X, Y, and S points as well as the symmetry lines along X–S–Y, in Fig. 5.8, bands are degenerate and the product of the parities of the wave functions is always even. In contrast, a product is either even or odd at the Γ point since there is no restriction of degeneracy. Thus, when one takes the product of parities from the lowest band, there is a band gap, as shown in Fig. 5.9c, and a material is insulating when the product is even. There can be crossing of bands with different parities between the Γ point and the zone boundary (Fig. 5.9b). In the latter case, a material is semimetal with a DNL, as drawn in Fig. 5.9a.

The presence of the DNL can be predicted by topology based on the Z_2 topological index, defined as

$$\mathbb{Z}_2 = \prod_{\text{occupied}} \Gamma_i(\sigma_z) = \begin{cases} 1 & (\text{even}: 0, 2, 4 \ldots) \\ -1 & (\text{odd}: 1, 3, 5 \ldots) \end{cases}$$

where σ_z is the mirror operation of an x–y plane and $\Gamma_i(\sigma_z)$ is the mirror eigenvalue of the ith band. The product is taken for all the occupied bands at the Γ point. The cases in Fig. 5.9b, c correspond to $Z_2 = -1$ and $Z_2 = 1$, respectively. When one applies this topological formula to borophane in the forms (5-7)-α_1-$B_{16}H_{16}$ and (5-7)-α_2 -$B_{16}H_{16}$, one finds the $Z_2 = -1$ for both cases, confirming the semimetallicity with DNLs in the band calculations (Fig. 5.8). Figure 5.9e, f presents other examples of borophane with the non-symmorphic space group, P_{bam} (layer group No. 44 or 3D space group No. 55), that allows application of the topological formula. These borophane layers result from the hydrogenation of borophene with the five-, six-, and seven-membered rings, Fig.5.9d. (5-6-7)-γ_1-borophane has $Z_2 = -1$, and

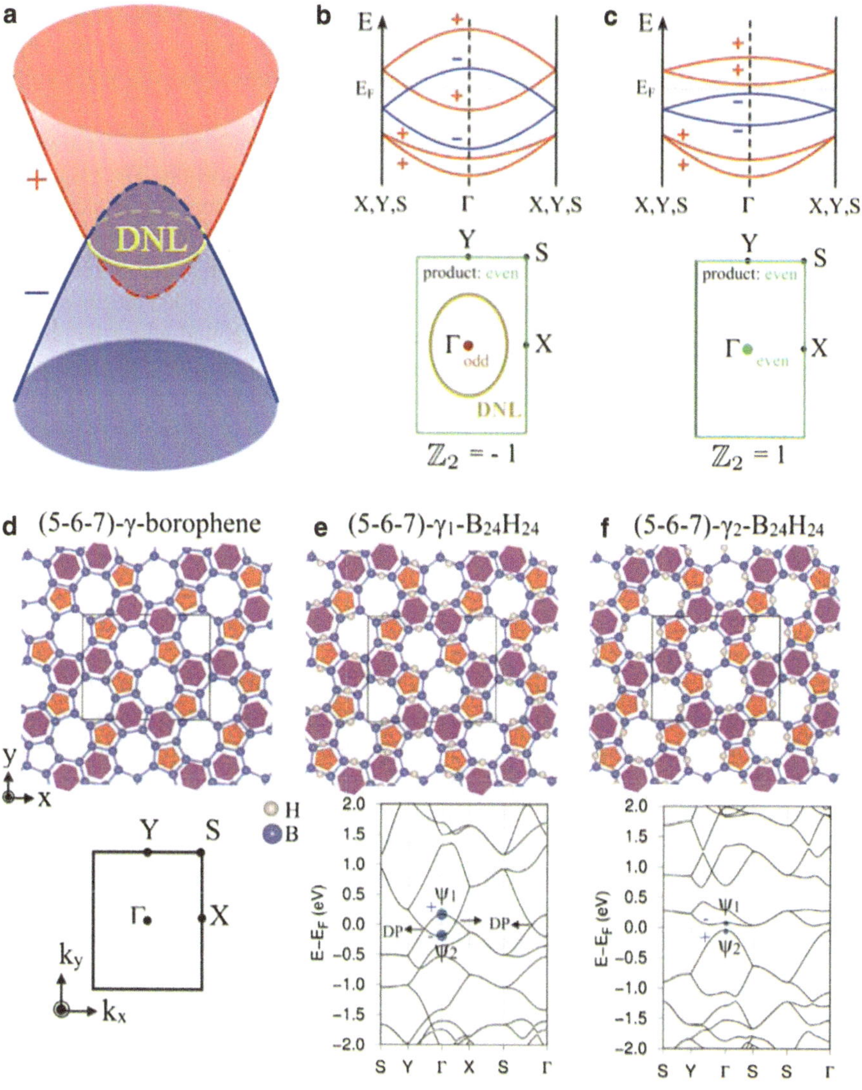

Fig. 5.9 (**a**) A drawing of the DNL. (**b**, **c**) Dispersion curves of electronic bands in the layer group No. 44. Odd and even symmetries of the electronic bands are labeled + and –, respectively. Appearance of DNL, depending on symmetries of the wave functions [78]. Structure models of monolayers of (**d**) (5-6-7)- γ-borophene, (**e**) (5-6-7)- γ_1-$B_{24}H_{24}$, and (**f**) (5-6-7)- γ_2-$B_{24}H_{24}$. Boron and hydrogen atoms are shown as blue and light pink balls, respectively. Locations of the five-membered and six-membered rings are indicated with red pentagons and purple hexagons, respectively. Parities of mirror reflection symmetry with respect to the x–y plane for the Ψ_1 and Ψ_2 bands are labeled with + and – signs, respectively

the calculated band structure is semimetal with a DNL (Fig. 5.9e), while (5-6-7)-γ_2-borophane has $Z_2 = 1$ and the calculated band structure is insulating (Fig. 5.9f). These calculation results justify the topological predictions of a DNL in the borophane layers.

Up to now, samples of borophane have been synthesized in powder form using the liquid exfoliation method, as described in the previous section. The electronic states of such samples can be directly probed using X-ray absorption spectroscopy (XAS) and X-ray emission spectroscopy (XES). In the optical absorption of a sample, there is an increase of the X-ray absorption at an energy that is specific to each element, followed by a spectral fine structure. The edge is called the absorption edge, and one can obtain information about the boron compound when the photon energy is tuned at $h\nu_{XAS} = 190$ eV for boron. When one focuses on the near-edge region ($\Delta h\nu_{XAS} = 20$–50 eV from the absorption edge), the spectroscopy is called near-edge X-ray absorption fine structure (NEXAFS) or X-ray absorption near-edge structure (XANES). NEXAFS spectra directly reflect the unoccupied electronic states of a sample. On the other hand, there is also an increase of the X-ray emission when the emission photon energy with matches that of the absorption edge, for example, about $h\nu_{XES} = 190$ eV for boron. The X-ray emission spectra directly reflect the occupied electronic states of a sample. Thus, a combination of the XAS and XES measurements unveils both the unoccupied and occupied electronic states of boron compound samples.

Figure 5.10 shows a collection of the B K-edge XAS/XES data of powder samples of the HB sheet, the $B(OH)_3$ sheet, the B_2O_3 grain, and the MgB_2 crystal. For comparison, the C K-edge XAS/XES spectra of the HOPG crystal are also shown. The insulating samples of $B(OH)_3$ and B_2O_3 apparently show energy gaps, while the XAS and XES spectral features of the metallic MgB_2 sample overlap. This spectral overlap is also found in the HB sheets. The appearance is the same as that of HOPG, which is the standard reference for a semimetallic sample. These data provide direct evidence that the HB sheet has a gapless electronic structure at the Fermi level, which is consistent with the band calculation result of the semimetallicity in Fig. 5.7.

5.1.4 Application of Borophane

C-borophane (Figs. 5.2c, 5.4a, and 5.3a) was studied theoretically for several applications such as batteries [40], sensors [41], and catalysts [42]. Specifically, when used in Li and Na ion batteries, superior anode properties were reported [40] by which the maximum Li content is estimated to be $Li_{0.445}B_2H_2$, which gives rise to a material with a maximum theoretical specific capacity of 504 mAh/g together with an average voltage of 0.43 V versus Li/Li^+. Likewise, for Na ion batteries, the maximum theoretical capacity and average voltage were estimated to be 504 mAh/g and 0.03 V versus Na/Na^+, respectively. C-Borophane has also been reported to be useful for sensing alcohol vapor using the band gap widening caused by alcohol

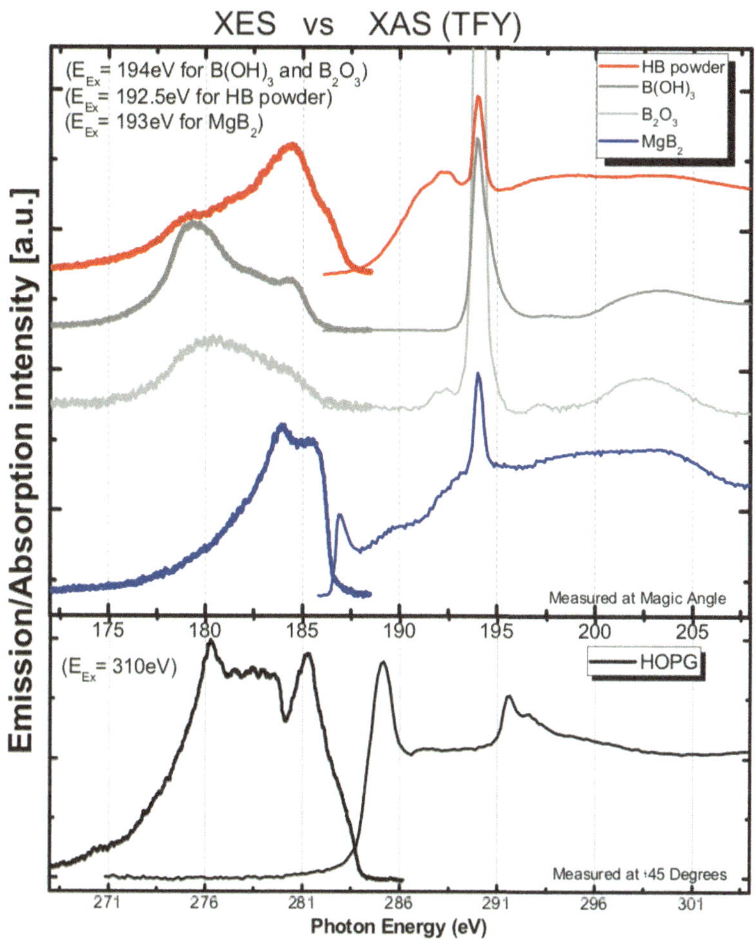

Fig. 5.10 XES and XAS spectra of powders of a HB sheet, $B(OH)_3$, B_2O_3, and MgB_2 at the B K-shell absorption edge. The XES measurements were made at the excitation energy (E_{EX}), indicated in the figure. For a comparison, XES and XAS spectra of HOPG, taken at C K-edge, are shown. (Reprinted by permission from Tateishi et al. [38]. Copyright (2019) by the American Physical Society)

adsorption on borophane [41]. The superior properties of C-borophane have also been reported through computational exploration of its application as a catalyst support material in the oxygen reduction reaction (ORR) and oxygen evolution reaction (OER) [42]. By combining defect induced C-borophane with single metal atoms of Fe or Co, the ORR electrocatalytic property is expected to be superior to the current most active catalyst Pt(111), and by combining with Rh, the OER electrocatalytic property is expected to be superior to the current most active catalyst RuO_2.

In the case of graphene-like borophane (hydrogen boride, HB sheet, Fig. 5.1), intriguing giant thermal conductance [43], controllable hydrogen release temperature by self-doping or charging (see section §5.1.1) [14, 15], reversible hydrogen storage ability by Li-decoration [44], excellent anode properties for ion batteries [45, 46], band engineering for electronic devices by applying mechanical loading [47, 48], and promising properties as a current limiter and photodetector [49] have been reported theoretically. Specifically, it exhibits a comparable lattice thermal conductance (4.07 nWK^{-1} nm^{-2}) to graphene (4.1 nWK^{-1} nm^{-2}) and an electron thermal conductance (3.6 nWK^{-1} nm^{-2}) which is almost ten times that of graphene. As a result, the total thermal conductance of 2D hydrogen boride is about twofold that of graphene and is the highest value in all known 2D materials. Moreover, tensile strain along the armchair direction (x direction in Fig. 5.1a) leads to an increase in carrier density, significantly increasing electron thermal conductance. The increase in electron thermal conductance offsets the reduction in phonon thermal conductance, contributing to an abnormal increase in thermal conductance [43]. It has also been reported that the HB sheet shows an unexpectedly high affinity for Li in comparison to other alkali and alkaline earth metals (Na, K, Ca, Mg, and Al) with a binding energy of −2.38 eV, which is much larger than its bulk cohesive energy (−1.63 eV). The energy barriers of Li diffusion on an HB sheet are determined to be around 1.12 eV, showing both high dynamic and thermodynamic stability without the issue of cluster formation. Moreover, a Li-decorated HB sheet is expected to achieve a high theoretical gravimetric density of 11.57 wt% with an average H_2 adsorption energy of −0.17 eV, demonstrating it holds great potential in massive hydrogen storage under ambient conditions [44]. As an anode battery, the HB sheet was also theoretically reported to lead to a high specific capacity of 861.78 mA h g^{-1} for Li ions [45]. In contrast, the H vacancies that emerged in the HB sheet electrodes were reported to improve both potassium ion intercalation and potassium ion hopping, yielding a high theoretical capacity (1138 mAh/g), which was among the highest reported value in the literature for K ion anode materials. Moreover, the HB sheet was reported to present an ultralow out-of-plane bending stiffness, rivaling that of graphene, which endows it with better flexibility to accommodate the repeated bending, rolling, and folding when applied in wearable devices. It also has a high in-plane stiffness (157 N/m along the armchair and 109 N/m along the zigzag, which are the x and y directions in Fig. 5.1a, respectively). The HB sheets can thus be used as a brand-new flexible and lightweight potassium ion battery anode material with high capacity, low voltage, and desirable rate performance [46]. The mechanical response of the HB sheet was reported to be anisotropic, with an elastic modulus of 131 N m^{-1} and a high tensile strength of 19.9 N m^{-1} along the armchair direction (x direction in Fig. 5.1a). By applying a mechanical loading, the metallic electronic character of the HB sheets can be altered to direct band gap semiconducting, which is very appealing for application in nanoelectronics [47].

The solid-acid catalytic property of graphene-like borophane (hydrogen boride, Figs. 5.1 and 5.11a) synthesized as a powder has been experimentally revealed [39] as well as its ability to release hydrogen molecules using ultraviolet-induced

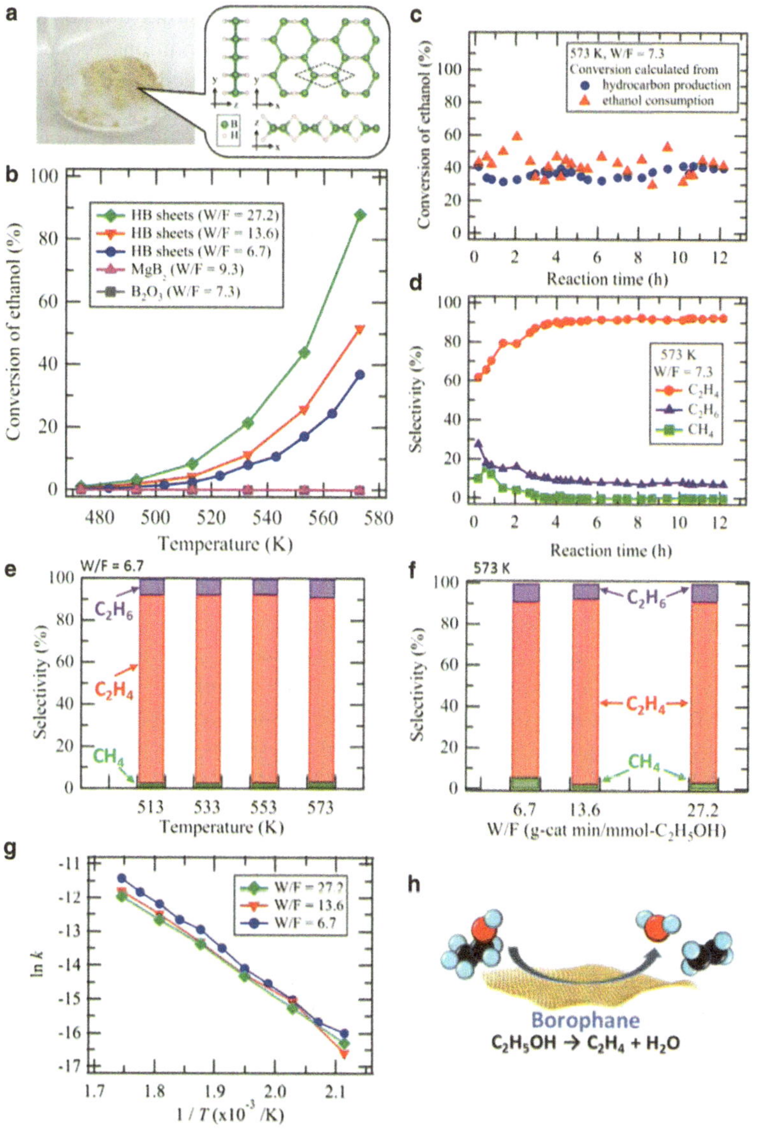

Fig. 5.11 Solid-acid catalytic property of graphene-like borophane [39]. (**a**) Photograph of HB sheets in powder form and schematic of the proposed local structure. (**b**) Conversion of ethanol vs. temperature. The results for HB, B_2O_3, and MgB_2 are plotted for various W/F (g min/mmol) conditions. (**c**) Conversion of ethanol on HB sheets as a function of reaction time at 573 K and W/F = 7.3 g min/mmol, estimated from hydrocarbon production (circles) and ethanol consumption (triangles). (**d**) Selectivity vs. reaction time. (**e**) Selectivity of ethanol reformation by HB as a function of temperature at W/F = 6.7 g·min/mmol. (**f**) Selectivity of ethanol reformation by HB at 573 K and W/F = 6.7, 13.6, and 27.2 g·min/mmol. (**g**) Arrhenius plot of the ethanol-reforming reaction rate in the presence of HB sheets for W/F = 27.2, 13.6, and 6.7 g·min/mmol. (**h**) Schematic image of the catalytic reaction

decomposition at room temperature [50]. Figure 5.11b shows the conversion of ethanol as a function of temperature at various W/F conditions (contact time); W/F (g min/mmol) is the weight of the catalyst (g) divided by the flow rate of C_2H_5OH (mmol/min). In every case, the sample was heated at 573 K under an Ar flow for 1 h prior to the measurement. It is clear that the HB sheets exhibit distinct ethanol conversion. The conversion remains constant for a long period (Fig. 5.11c). In the case shown in Fig. 5.11c, the total amount of hydrocarbons at 13.4 h is 3.2 mmol, which is approximately half of the total number of B atoms used in the HB sheets. The conversion estimated from the ethanol consumption (Fig. 5.11c) and that from the hydrocarbon production are almost the same at 12 h (approximately 40%), indicating that the hydrocarbon product does not accumulate on the catalyst (HB sheets) during the steady-state catalytic process. Hence, these results indicate that the HB sheets catalytically convert ethanol. The catalytically converted products were predominantly ethylene (C_2H_4) and water as shown in Fig. 5.11d. The same characteristics were observed at all of the measured temperatures and W/F conditions, as shown in Fig. 5.11e, f; that is, C_2H_4 was always the main product, and the selectivity for the total hydrocarbon obtained was almost the same. These results indicate that the major catalytic reaction of ethanol reformation by the HB sheets is the dehydration reaction: $C_2H_5OH \rightarrow C_2H_4 + H_2O$ (Fig. 5.11h). It is known that if a catalyst promotes the dehydration reaction of ethanol, it is a solid acid; however, if the dehydrogenation reaction occurs, producing an acetaldehyde, the catalyst is a base catalyst. The HB sheets are therefore solid-acid catalysts. Figure 5.11g shows the Arrhenius plot of the ethanol-reforming reaction rate k, which was calculated by assuming a first-order reaction, for the results obtained using the HB sheets under various W/F conditions. As the obtained linear plots are almost the same and are independent of W/F, the reaction can be (at least apparently) considered a first-order reaction. From the slope and the intersect, the apparent activation energy E_a and pre-exponential factor A were estimated to be 102.8 ± 5.5 kJ/mol and 3.5×10^4 s^{-1}, respectively. The derived E_a is comparable to the reported activation energies for the catalytic dehydration of ethanol over Al_2O_3 (53–155 kJ/mol), silica–alumina (125.5 kJ/mol), microporous Fe-ZSM-5 (137.7–271.1 kJ/mol), and the Lewis acidic Zr-KIT-6 catalyst (79 kJ/mol). The formation rate of ethylene on the HB catalyst at 573 K and W/F = 7.3 g·min/mmol was 2.4 ± 0.1 mmol/g·h (Fig. 5.11c). Chen and co-workers reported that the formation rate of ethylene from ethanol using the commercial SynDol (Al_2O_3–MgO/SiO_2) catalyst was 7.8 mmol/g·h at 591 K, with a weight hourly space velocity (WHSV) of 0.23 h^{-1} in a fixed-bed reactor (this value is calculated using the reported yield of ethylene, which is 0.22 g/gcat·h). They also reported a higher formation rate of 535 mmol/g·h on Ti/γ- Al_2O_3 at 633 K and a WHSV of 26 h^{-1} using the microreactor (calculated using a value of 15 g/gcat·h). The formation rate of the HB sheets is, thus, lower than that of state-of-the-art catalysts but is in the same order as that of the commercial SynDol catalyst.

5.2 Boron Compound Sheets

5.2.1 Boron Nitride Sheets

Boron nitride (BN) is a solid chemical compound of boron and nitrogen that has been used as an electric insulator, a heat resistor, and in the cutting tool industry [51–54, 58]. There are two well-known crystalline forms, hexagonal boron nitride (h-BN) and cubic boron nitride (c-BN). A crystal of h-BN has stacked BN planes, Fig. 5.12, with in-plane sp^2 covalent bond is, just like graphene, and has been called "white graphite" [55]. The calculated band structure of a h-BN sheet is shown in Fig. 5.12 [56, 57]. The layer is insulating.

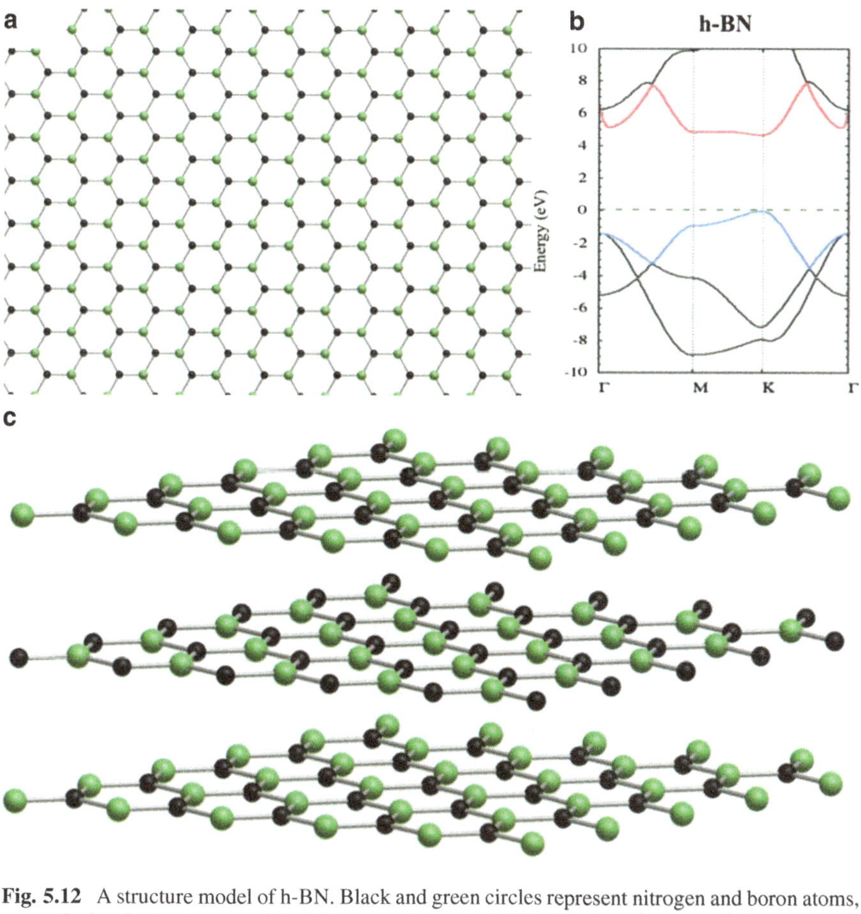

Fig. 5.12 A structure model of h-BN. Black and green circles represent nitrogen and boron atoms, respectively. A structure model of layer stacking in h-BN. Black and green circles represent nitrogen and boron atoms, respectively. Calculated band diagram of a h-BN layer. (Reprinted by permission from Miró et al. [57]. Copyright (2014) ROYAL SOCIETY OF CHEMISTRY)

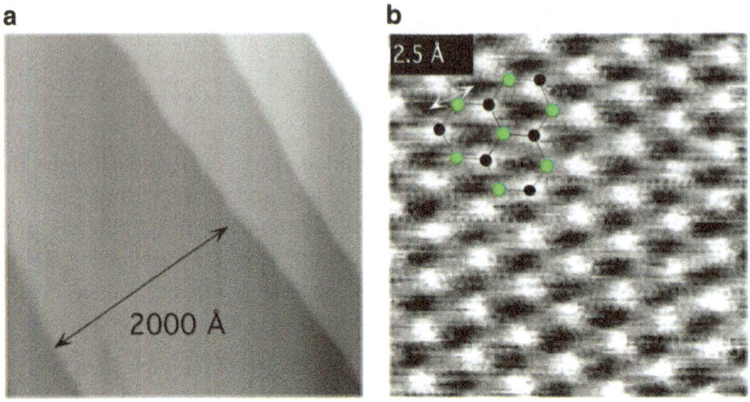

Fig. 5.13 STM images of h-BN on Ni(111): (**a**) wide and (**b**) local area. In (**b**), the structure model of h-BN is also shown [60, 81–83]. Black and green circles represent nitrogen and boron atoms, respectively. (By courtesy of Thomas Greber)

Atomic sheets of h-BN can be synthesized via two routes: top-down exfoliation and bottom-up growth. There are a variety of exfoliation methods including mechanical and chemical [53, 54]. It is worth mentioning that a high-quality 3D crystal of BN is achieved by growth under high pressures (GPa) using solvents of metal-BN such as Ba-BN, in Japan, and is delivered to frontier laboratories all over the world by Takashi Taniguchi and Kenji Watanabe [59]. Mechanical exfoliation, using Scotch tape or a ball milling process, is simple but significantly limits the flake size. Chemical exfoliation is essentially a process of liquid-phase exfoliation that disperses the h-BN sheets in organic solvents, such as chloroform and DMF, with the assistance of sonication. This quick and easy approach leads to high yields but suffers from surface contamination and small flake size. The limited sample size using exfoliation methods currently hinders application in large surface-area devices.

A route using bottom-up growth is carried out by self-assembly of ordered h-BN sheets on specific substrates using deposition methods, in particular using a chemical vapor deposition (CVD) method. Mono- or few-layer 2D h-BN grows through decomposition of boron-based gas molecules, such as borazine ($B_3N_3H_6$) and trichloroborazine ($B_3N_3H_3Cl_3$), on transition metal surfaces, such as Ni, Pd, Ru, Pt, or Rh. The CVD growth method has the advantage of synthesizing a wide area of h-BN sheets that are high-quality [53, 54]. A typical scanning tunneling microscope (STM) image is shown in Fig. 5.13 [60, 81–83]. However, it also uses catalytic substrates at high temperatures and requires transfer of the processes to other substrates for application. These restrictions currently require improvement before construction of high-performance devices can be achieved.

An atomic sheet of h-BN is chemically inert, mechanically robust, thermally conductive, and electrically insulating (a band gap of 6 eV). Because of these unique properties, it has attracted significant interest in the study of 2D materials

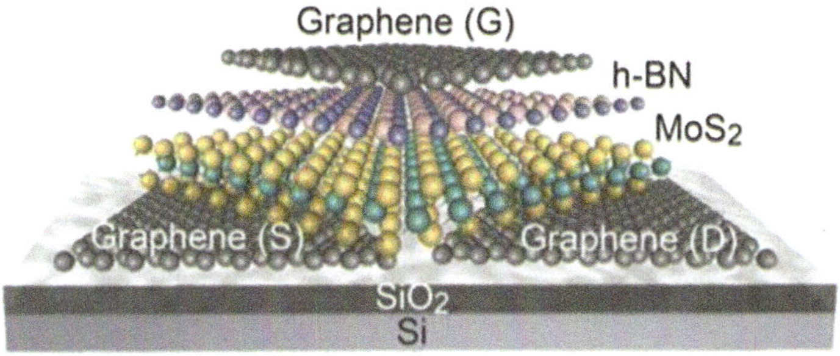

Fig. 5.14 Design of an all monoatomic (monatomic) sheet FET transistor with a layer of h-BN as the insulator. (Reprinted by permission from Roy et al. [61]. Copyright (2014) American Chemical Society)

and its technical application as a component in atomic or nano-devices has been explored. For example, in combination with layers of graphene and MoS_2, the h-BN sheet can act as an insulating layer building block in a transistor with a van der Waals heterostructure [61], Fig. 5.14. Applications of such 2D electronics have been widely explored, such as memory devices, capacitors, filters, and sensors. The layer on metal substrates are also versatile templates for atoms, molecules, and nanostructures [84]. New research fields are still on the rise today and sheets of h-BN are expected to be a key material for these new developments.

5.2.2 *Boron Carbon Nitride Sheets*

In the graphenic 2D systems, a layer of graphene (honeycomb lattice of C) is a zero-band gap semiconductor and that of h-BN is an insulator. A semiconducting layer has been investigated for graphenic BCN, which is a ternary network based on atoms of boron, carbon, and nitrogen that occupy the graphenic sites in a 2D sheet [62]. Theoretical research has predicted various models of such 2D BCN compounds and reported a variety of properties, which depend on the chemical stoichiometry and the atomic arrangement of B, C, and N [63–67]. There have also been many attempts of experimental realization [68–72]. Figure 5.15 shows an example of the atomic model that has been found to be stable using first-principles calculations and that to be realized on Ir(111) [62]. The three neighboring B, N, and C atoms take the configurations of, B_2NC, BN_2C, or BNC_2. The band calculation shows that the 2D BCN layer is semiconducting with a direct band gap of 1–1.5 eV and is dependent on the DFT functional (Fig. 5.15) [62]. These results demonstrate that the electronic structure of BCN is intermediate to those of graphene and h-BN. Since sheets of graphene and h-BN have been adopted as conductive and insulating layers, respectively, those of h -BCN are a promising semiconducting layer for 2D electronic materials.

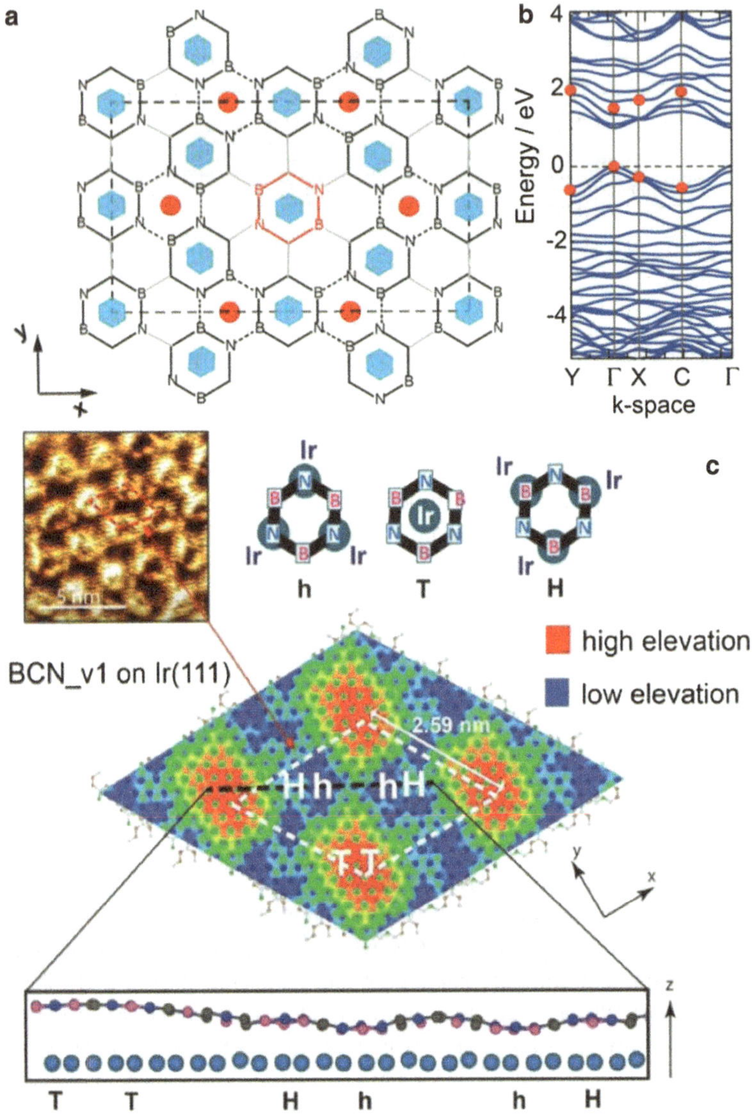

Fig. 5.15 (**a**) A possible h-BCN model, BCN_v1, that has a boding network of B−C, B−N, and C−N, shown as the precursor molecule [62]. The red circles and blue hexagons indicate units with different compositions of B, C, and N atoms. (**b**) The electronic band structures of a freestanding sheet (BCN_v1), calculated using density functional theory with the two types of functional, blue curves, and red points. The zero energy corresponds to the Fermi level [62]. (**c**) STM images of a BCN layer on Ir(111) and a comparison with the BCN_v1 model with its structural corrugation. (Reprinted by permission from Beniwal et al. [62]. Copyright (2017) American Chemical Society)

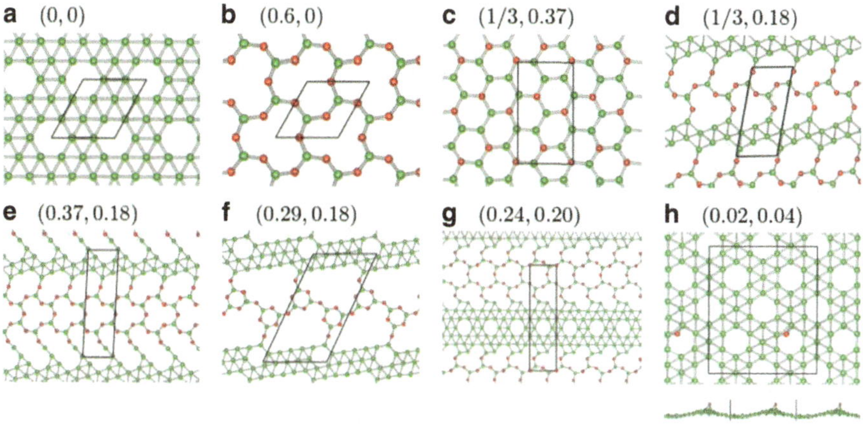

Fig. 5.16 Structure models of $B_{1\text{-}x}O_x$ sheets with different tuples of oxygen fraction, x, and mixing energy, E_{mix}, as indicated in the parentheses. Boron and oxygen atoms are indicated by green and red balls, respectively [74]. (By courtesy of Jens Kunstmann)

5.2.3 Boron Oxide Sheets

Boron has a high chemical affinity for oxygen, and there are various compounds of boron oxides, such as boron trioxide (B_2O_3) and boron suboxide (B_6O). Layers of borophene have also been reported to be easily oxidized, but little is known about the oxidizations or properties of 2D boron oxides (2D BO). Currently, theoretical research has been performed for sheets of B_xO (x = 4 ~ 8) [73] and $B_{1\text{-}x}O_x$ (x = 0 ~ 0.6), where B_2O_3 at x = 0.6 [74].

In the case of 2D boron oxide, $B_{1\text{-}x}O_x$, stability is examined theoretically from borophene (α'-sheet) at x = 0 to B_2O_3 at x = 0.6 [74]. As shown in Fig. 5.16a, boron atoms in borophene have high in-plane coordination numbers (N_{coord} = 5 and 6). On the other hand, N_{coord} = 3 for boron atoms, while N_{coord} = 2 for oxygen atoms in Fig. 5.16b. The first-principles calculations of the 149 boron-oxygen system shows that there are only stable freestanding layers of borophene and B_2O_3. At the intermediate composition, the 2D layer is composed of a heterogeneous mixture, as shown in Fig. 5.16c–g. These findings indicate that oxygen atoms prefer twofold coordination in 2D boron oxide.

Predictions of 2D boron suboxides have also been performed using a global search based on first-principles calculations [73]. Figure 5.17 shows a collection of the lowest-energy structures of 2D boron oxide with different B/O ratios, B/O = 4 ~ 8. All of them have a nonplanar structure. Layers of B_5O, B_7O, and B_8O consist of boron ribbons that are connected to each other by B–O–B bonds. These layers have triangular lattices and hexagonal holes. In the B_5O and B_7O layers, there are hexagonal holes with B and O atoms at the edges, while in B_8O there are additional hexagonal holes with only B edges. In B_4O, boron ribbons are connected by the B–O square lattice and there are hexagonal holes with B and O edges. 2D

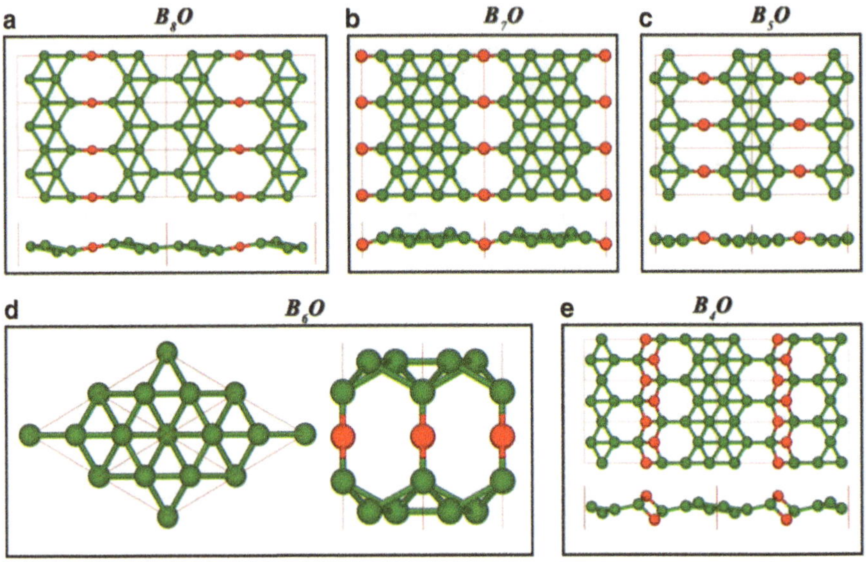

Fig. 5.17 Structure models of suboxide sheets, B_xO, with different x values. Boron and oxygen atoms are indicated by green and red balls, respectively. (Reprinted by permission from Zhang et al. [73]. Copyright (2017) American Chemical Society)

crystals of B_6O have quite a distinctive structure; they are composed of two boron layers of the buckled triangular lattices that sandwich a layer of oxygen atoms. The authors of the paper [73] mentioned that, for B/O = 5 ~ 8, 2D boron oxide can be achieved by insertion of oxygen atoms between the B–B bonds, while, for B/O = 4, the tetrahedral configuration shows possible bonding of one O atom with three B atoms. This tendency indicates variations in the chemical coordination number of O atoms are caused by the chemical stoichiometry. Calculated band structures of 2D boron oxides show that B_5O, B_7O, and B_8O are metallic. In the case of B_4O, the layer is semiconducting with anisotropic band edges. The strong optical anisotropy of B_4O may be applied in 2D polarization-sensitive photodetectors. B_6O is a 2D DNL semimetal, as shown in Fig. 5.18. Recently, a B_2O monolayer of honeycomb borophene oxide was proposed as illustrated in Fig. 5.19 [75]. The calculation shows that the electronic band structure is composed of a nodal loop centered around the symmetry point that can be characterized by the nontrivial topological property (Fig. 5.19). Since 2D DNL semimetals are rarely studied, these electronic states of the unique Dirac ring in B_6O and B_2O may show intriguing quantum transport phenomena. It is of note that the existence of a crystal of B_2O has been debated; however, the possible synthesis of honeycomb B_2O is proposed by oxidization of honeycomb borophene on Al(111). Layer growth on substrates often results in 2D boron layers that do not exist in nature in the freestanding manner. There have been varieties of borophene reported on various substrates that can be used as starting materials to synthesize the novel boron oxide or suboxide layers.

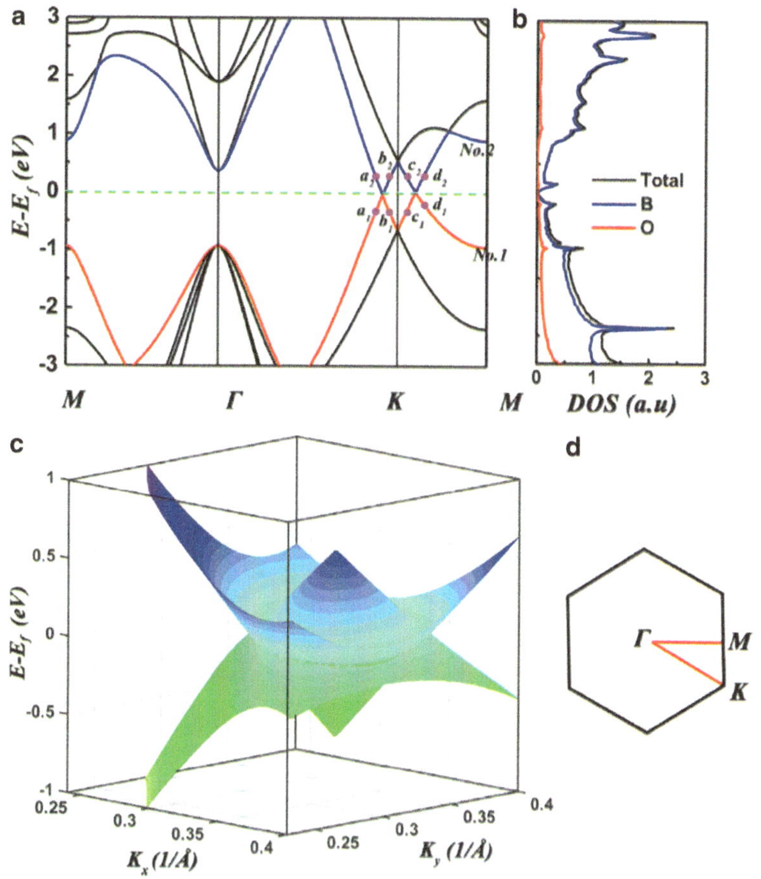

Fig. 5.18 (**a**) Calculated band structure and (**b**) optical absorption spectra with two types of polarization for the B_4O model [73]. (**a**) Calculated band diagram and (**b**) density of states of the B_6O model [73]. (**c**) Band dispersion curves around the Fermi level. (**d**) The 2D Brillouin zone. (Reprinted by permission from Zhang et al. [73]. Copyright (2017) American Chemical Society)

5.3 Summary and Outlook

In this chapter, we present the theoretical predictions of various chemically modified borophene polymorphs and note several intriguing physical and chemical properties that have been reported, including superiority as battery materials, application as hydrogen storage materials, and electronic and/or mechanical applications (such as memory, sensor, and/or high hardness materials). There are other predictions of the polymorphs for 2D boron phosphides [76] and boron chalcogenides [79, 80]. The experimental synthesis of hydrogenated borophene in powder form has only recently been reported, in which a top-down synthesis method employing ion-exchange of protons with Mg cations in MgB_2 was used. The method used for

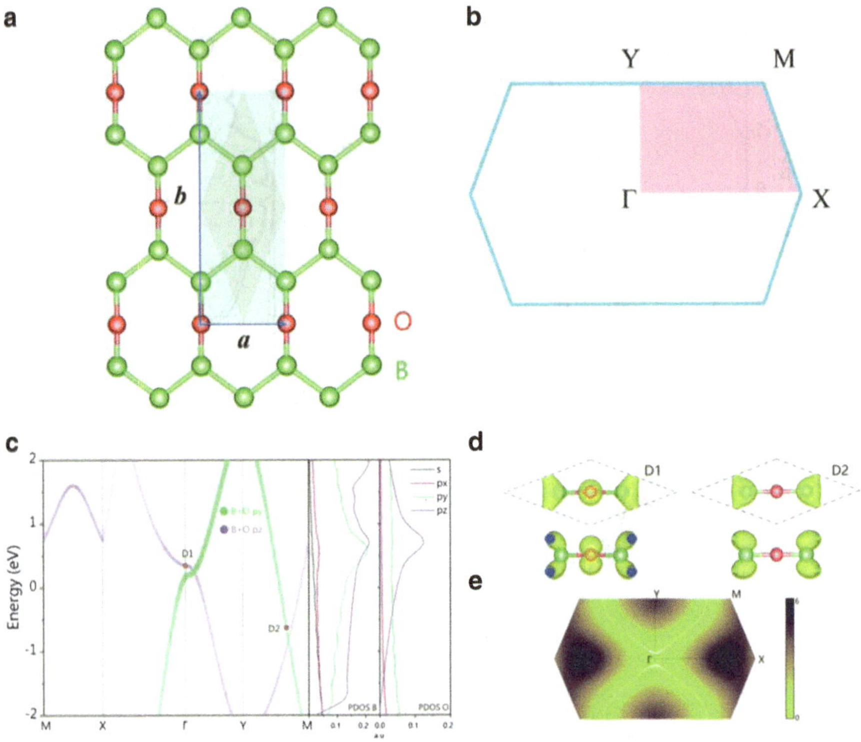

Fig. 5.19 (**a**) Structure models of sheets of h (hexagonal)-B_2O. Boron and oxygen atoms are indicated by green and red balls, respectively. The conventional and primitive cells are shown in the blue (rectangular) and orange (rhombus) areas, respectively. (**b**) The 2D Brillouin zone. The irreducible Brillouin zone is colored in red [75]. (**c**) Calculated band structure of the h-B_2O model. (**d**) Charge densities at D1 and D2 in (**c**). (**e**) The energy difference of the B + O p_y and B + O p_z bands around the Γ point. The white curves indicate the location of the nodal loop. (Reprinted by permission from Zhong et al. [75]. Copyright (2019) ROYAL SOCIETY OF CHEMISTRY)

this synthesis can be categorized as an alternative route for the production of 2D boron materials [20, 22], which may open new avenues to achieve an increase in the number of chemically modified borophenes. A recent report for the preparation of freestanding borophene may further accelerate such a top-down approach [77] to prepare other types of chemically modified borophene. Simultaneously, we can expect the appearance of well-defined chemically modified borophene sheets produced using a bottom-up approach through the modification of borophene sheets prepared on a metal substrate using evaporation in an ultra-high vacuum. As in the case of graphene, both bottom-up and top-down approaches are required in order to apply the material in a wide variety of fields.

References

1. S.Z. Butler, S.M. Hollen, L. Cao, Y. Cui, J.A. Gupta, H.R. Gutiérrez, T.F. Heinz, S.S. Hong, J. Huang, A.F. Ismach, et al., Progress, challenges, and opportunities in two-dimensional materials beyond graphene. ACS Nano **7**(4), 2898–2926 (2013). https://doi.org/10.1021/nn400280c
2. M. Osada, T. Sasaki, Two-dimensional dielectric nanosheets: Novel nanoelectronics from nanocrystal building blocks. Adv. Mater. **24**(2), 210–228 (2012). https://doi.org/10.1002/adma.201103241
3. D. Deng, K.S. Novoselov, Q. Fu, N. Zheng, Z. Tian, X. Bao, Catalysis with two-dimensional materials and their heterostructures. Nat. Nanotechnol. **11**(3), 218–230 (2016). https://doi.org/10.1038/nnano.2015.340
4. Z. Zhang, E.S. Penev, B.I. Yakobson, Two-dimensional boron: Structures, properties and applications. Chem. Soc. Rev. **46**(22), 6746–6763 (2017). https://doi.org/10.1039/c7cs00261k
5. T. Kondo, Recent progress in boron nanomaterials. Sci. Technol. Adv. Mater. **18**(1), 780–804 (2017). https://doi.org/10.1080/14686996.2017.1379856
6. Y. Jiao, F. Ma, J. Bell, A. Bilic, A. Du, Two-dimensional boron hydride sheets: High stability, massless Dirac fermions, and excellent mechanical properties. Angew. Chemie – Int Ed. **55**(35), 10292–10295 (2016). https://doi.org/10.1002/anie.201604369
7. A.R. Oganov, V.L. Solozhenko, Boron: A hunt for superhard polymorphs. J. Superhard Mater. **31**(5), 285–291 (2009). https://doi.org/10.3103/S1063457609050013
8. A.J. Mannix, B. Kiraly, M.C. Hersam, N.P. Guisinger, Synthesis and chemistry of elemental 2D materials. Nat. Rev. Chem. **1**, 0014 (2017). https://doi.org/10.1038/s41570-016-0014
9. S.Y. Xie, Y. Wang, X.B. Li, Flat boron: A new cousin of graphene. Adv. Mater. **1900392**, 1–13 (2019). https://doi.org/10.1002/adma.201900392
10. I. Boustani, New quasi-planar surfaces of bare boron. Surf. Sci. **370**(2–3), 355–363 (1997). https://doi.org/10.1016/S0039-6028(96)00969-7
11. E.S. Penev, S. Bhowmick, A. Sadrzadeh, B.I. Yakobson, Polymorphism of two-dimensional boron. Nano Lett. **12**(5), 2441–2445 (2012). https://doi.org/10.1021/nl3004754
12. X. Wu, J. Dai, Y. Zhao, Z. Zhuo, J. Yang, X.C. Zeng, Two-dimensional boron monolayer sheets. ACS Nano **6**(8), 7443–7453 (2012). https://doi.org/10.1021/nn302696v
13. A.K. Geim, I.V. Van Der Grigorieva, Waals heterostructures. Nature **499**(7459), 419–425 (2013). https://doi.org/10.1038/nature12385
14. T.A. Abtew, B.C. Shih, P. Dev, V.H. Crespi, P. Zhang, Prediction of a multicenter-bonded solid boron hydride for hydrogen storage. Phys. Rev. B - Condens. Matter Mater. Phys. **83**(9), 094108 (2011). https://doi.org/10.1103/PhysRevB.83.094108
15. T.A. Abtew, P. Zhang, Charging-assisted hydrogen release mechanism in layered boron hydride. Phys. Rev. B - Condens. Matter Mater. Phys. **84**(9), 1–6 (2011). https://doi.org/10.1103/PhysRevB.84.094303
16. Z.Z. Wang, T.Y. Lü, H.Q. Wang, Y.P. Feng, J.C. Zheng, New crystal structure prediction of fully hydrogenated borophene by first principles calculations. Sci. Rep. **7**, 1–11 (2017). https://doi.org/10.1038/s41598-017-00667-x
17. W.-L. Li, C. Romanescu, T. Jian, L.-S. Wang, Elongation of planar boron clusters by hydrogenation: Boron analogues of polyenes. J. Am. Chem. Soc. **134**(32), 13228–13231 (2012). https://doi.org/10.1021/ja305744a
18. C.H. Hu, A.R. Oganov, Q. Zhu, G.R. Qian, G. Frapper, A.O. Lyakhov, H.Y. Zhou, Pressure-induced stabilization and insulator-superconductor transition of BH. Phys. Rev. Lett. **110**(16), 1–5 (2013). https://doi.org/10.1103/PhysRevLett.110.165504
19. H. Nishino, T. Fujita, N.T. Cuong, S. Tominaka, M. Miyauchi, S. Iimura, A. Hirata, N. Umezawa, S. Okada, E. Nishibori, et al., Formation and characterization of hydrogen boride sheets derived from MgB_2 by cation exchange. J. Am. Chem. Soc. **139**(39), 13761–13769 (2017). https://doi.org/10.1021/jacs.7b06153

20. G. Ozin, T. Siler, Catalyst: New materials discovery: Machine-enhanced human creativity. Chem **4**(6), 1183–1189 (2018). https://doi.org/10.1016/j.chempr.2018.05.011
21. S. Tominaka, R. Ishibiki, A. Fujino, K. Kawakami, K. Ohara, T. Masuda, I. Matsuda, H. Hosono, T. Kondo, Geometrical frustration of B-H bonds in layered hydrogen borides accessible by soft chemistry. Chem **6**, 406–418 (2020). https://doi.org/10.1016/j.chempr.2019.11.006
22. X.-M. Chen, X. Chen, Chemical syntheses of two-dimensional boron materials. Chem **6**, 324–326 (2020). https://doi.org/10.1016/j.chempr.2020.01.001
23. I. Boustani, Systematic ab initio investigation of bare boron clusters: Determination of the geometry and electronic structures of B_n (n=2–14). Phys. Rev. B **55**(24), 16426–16438 (1997). https://doi.org/10.1103/PhysRevB.55.16426
24. L.C. Xu, A. Du, L. Kou, Hydrogenated borophene as a stable two-dimensional Dirac material with an ultrahigh fermi velocity. Phys. Chem. Chem. Phys. **18**(39), 27284–27289 (2016). https://doi.org/10.1039/c6cp05405f
25. J.E. Padilha, R.H. Miwa, A. Fazzio, Directional dependence of the electronic and transport properties of 2D borophene and borophane. Phys. Chem. Chem. Phys. **18**(36), 25491–25496 (2016). https://doi.org/10.1039/C6CP05092A
26. L. Kou, Y. Ma, C. Tang, Z. Sun, A. Du, C. Chen, Auxetic and ferroelastic borophane: A novel 2D material with negative possion's ratio and switchable Dirac transport channels. Nano Lett. **16**(12), acs.nanolett.6b04180 (2016). https://doi.org/10.1021/acs.nanolett.6b04180
27. Z. Wang, T.-Y. Lü, H.-Q. Wang, Y.P. Feng, J.-C. Zheng, High anisotropy of fully hydrogenated borophene. Phys. Chem. Chem. Phys. **18**(46), 31424–31430 (2016). https://doi.org/10.1039/c6cp06164h
28. G. Liu, H.H. Wang, Y. Gao, J. Zhou, H.H. Wang, Anisotropic intrinsic lattice thermal conductivity of borophane from first-principles calculations. Phys. Chem. Chem. Phys. **19**(4), 2843–2849 (2016). https://doi.org/10.1039/C6CP07367K
29. M. Nakhaee, S.A. Ketabi, F.M. Peeters, Tight-binding model for borophene and borophane. Phys. Rev. B **97**(12), 125424 (2018). https://doi.org/10.1103/PhysRevB.97.125424
30. H. Nishino, T. Fujita, A. Yamamoto, T. Fujimori, A. Fujino, S.I. Ito, J. Nakamura, H. Hosono, T. Kondo, Formation mechanism of boron-based nanosheet through the reaction of MgB_2 with water. J. Phys. Chem. C **121**(19), 10587–10593 (2017). https://doi.org/10.1021/acs.jpcc.7b02348
31. K.B. Garg, T. Chatterji, S. Dalela, M. Heinonnen, J. Leiro, B. Dalela, R.K. Singhal, Core level photoemission study of polycrystalline MgB_2. Solid State Commun. **131**(5), 343–347 (2004). https://doi.org/10.1016/j.ssc.2004.01.006
32. A. Talapatra, S.K.K. Bandyopadhyay, P. Sen, P. Barat, S. Mukherjee, M. Mukherjee, A. Talapatra, S.K.K. Bandyopadhyay, P. Sen, P. Barat, et al., X-ray photoelectron spectroscopy studies of MgB_2 for valence state of Mg. Phys. C-Superconduct. Appl. **419**(3–4), 141–147 (2005). https://doi.org/10.1016/j.physc.2005.01.001
33. E.Z. Kurmaev, I.I. Lyakhovskaya, J. Kortus, A. Moewes, N. Miyata, M. Demeter, M. Neumann, M. Yanagihara, M. Watanabe, T. Muranaka, et al., Electronic structure of MgB_2 : X-ray emission and absorption studies. Phys. Rev. B **65**(13) (2002). https://doi.org/10.1103/PhysRevB.65.134509
34. L. Shao, X. Duan, Y. Li, Q. Yuan, B. Gao, H. Ye, P. Ding, A theoretical study of several fully hydrogenated borophenes. Phys. Chem. Chem. Phys. **21**(14), 7630–7634 (2019). https://doi.org/10.1039/c9cp00468h
35. S. Izadi Vishkayi, M. Bagheri Tagani, Current-voltage characteristics of borophene and borophane sheets. Phys. Chem. Chem. Phys. **19**(32), 21461–21466 (2017). https://doi.org/10.1039/c7cp03873a
36. P. Zhang, X.D. Li, C.H. Hu, S.Q. Wu, Z.Z. Zhu, First-principles studies of the hydrogenation effects in silicene sheets. Phys. Lett. Sect. A Gen. At. Solid State Phys. **376**(14), 1230–1233 (2012). https://doi.org/10.1016/j.physleta.2012.02.030
37. J.O. Sofo, A.S. Chaudhari, G.D. Barber, Graphane: A two-dimensional hydrocarbon. Phys. Rev. B - Condens. Matter Mater. Phys. **75**(15), 1–4 (2007). https://doi.org/10.1103/PhysRevB.75.153401

38. I. Tateishi, N.T. Cuong, C.A.S. Moura, M. Cameau, R. Ishibiki, A. Fujino, S. Okada, A. Yamamoto, M. Araki, S. Ito, et al., Semimetallicity of free-standing hydrogenated monolayer boron from MgB_2. Phys. Rev. Mater. **3**(2), 1–8 (2019). https://doi.org/10.1103/PhysRevMaterials.3.024004
39. A. Fujino, S. Ito, T. Goto, R. Ishibiki, J.N. Kondo, T. Fujitani, J. Nakamura, H. Hosono, T. Kondo, Hydrogenated borophene shows catalytic activity as solid acid. ACS Omega **4**(9), 14100–14104 (2019). https://doi.org/10.1021/acsomega.9b02020
40. N.K. Jena, R.B. Araujo, V. Shukla, R. Ahuja, Borophane as a benchmate of graphene: A potential 2D material for anode of Li and Na-ion batteries. ACS Appl. Mater. Interfaces **9**(19), 16148–16158 (2017). https://doi.org/10.1021/acsami.7b01421
41. V. Nagarajan, R. Chandiramouli, Sensing properties of monolayer borophane nanosheet towards alcohol vapors: A first-principles study. J. Mol. Graph. Model. **73**, 208–216 (2017). https://doi.org/10.1016/j.jmgm.2017.02.003
42. Y. Singh, S. Back, Y. Jung, Computational exploration of borophane-supported single transition metal atoms as potential oxygen reduction and evolution electrocatalysts. Phys. Chem. Chem. Phys. **20**(32), 21095–21104 (2018). https://doi.org/10.1039/c8cp03130d
43. J. He, D. Li, Y. Ying, C. Feng, J. He, C. Zhong, H. Zhou, P. Zhou, G. Zhang, Orbitally driven giant thermal conductance associated with abnormal strain dependence in hydrogenated graphene-like borophene. NPJ Comput. Mater. **5**(1), 1–8 (2019). https://doi.org/10.1038/s41524-019-0183-2
44. L. Chen, X. Chen, C. Duan, Y. Huang, Q. Zhang, B. Xiao, Reversible hydrogen storage in pristine and Li decorated 2D boron hydride. Phys. Chem. Chem. Phys. **20**(48), 30304–30311 (2018). https://doi.org/10.1039/C8CP05846F
45. V. Shukla, R.B. Araujo, N.K. Jena, R. Ahuja, Borophene's tryst with stability: Exploring 2D hydrogen boride as an electrode for rechargeable batteries. Phys. Chem. Chem. Phys. **20**(34), 22008–22016 (2018). https://doi.org/10.1039/c8cp03686a
46. P. Xiang, X. Chen, B. Xiao, Z.M. Wang, Highly flexible hydrogen boride monolayers as potassium-ion battery anodes for wearable electronics. ACS Appl. Mater. Interfaces **11**(8), 8115–8125 (2019). https://doi.org/10.1021/acsami.8b22214
47. B. Lei, Y.Y. Zhang, S.X. Du, Band engineering of B_2H_2 nanoribbons. Chinese Phys. B **28**(4) (2019). https://doi.org/10.1088/1674-1056/28/4/046803
48. B. Mortazavi, M. Makaremi, M. Shahrokhi, M. Raeisi, C.V. Singh, T. Rabczuk, L.F.C. Pereira, Borophene hydride: A stiff 2D material with high thermal conductivity and attractive optical and electronic properties. Nanoscale **10**, 3759–3768 (2018). https://doi.org/10.1039/c7nr08725j
49. Y. An, Y. Hou, H. Wang, J. Li, R. Wu, T. Wang, H. Da, J. Jiao, Unveiling the electric-current-limiting and photodetection effect in two-dimensional hydrogenated borophene. Phys. Rev. Appl. **11**(6), 1 (2019). https://doi.org/10.1103/PhysRevApplied.11.064031
50. R. Kawamura, N.T. Cuong, T. Fujita, R. Ishibiki, T. Hirabayashi, A. Yamaguchi, I. Matsuda, S. Okada, T. Kondo, M. Miyauchi, Photoinduced hydrogen release from hydrogen boride sheets. Nat. Commun. **10**, 4880 (2019). https://doi.org/10.1038/s41467-019-12903-1
51. E. Moran, *Boron Nitride: Properties, Synthesis and Applications (Chemistry Research Application)* (Nova Science Pub Inc, 2017)
52. Y.(.I.). Chen, *Nanotubes and Nanosheets: Functionalization and Applications of Boron Nitride and Other Nanomaterials* (CRC Press, 2015)
53. K. Zhang, Y. Feng, F. Wang, Z. Yang, J. Wang, Two dimensional hexagonal boron nitride (2D-HBN): Synthesis, properties and applications. J. Mater. Chem. C **5**(46), 11992–12022 (2017). https://doi.org/10.1039/c7tc04300g
54. H. Wang, Y. Zhao, Y. Xie, X. Ma, X. Zhang, Recent progress in synthesis of two-dimensional hexagonal boron nitride. J. Semicond. **38**, 031003 (2017)
55. E.A. Smith, Graphite and boron nitride ("white graphite"): Aspects of structure, powder size, powder shape, and purity. Powder Metall. **14**, 110–123 (1971)

56. A. Nagashima, N. Tejima, Y. Gamou, T. Kawai, C. Oshima, Electronic structure of monolayer hexagonal boron nitride physisorbed on metal surfaces. Phys. Rev. Lett. **75**(21), 3918–3921 (1995). https://doi.org/10.1103/PhysRevLett.75.3918
57. P. Miró, M. Audiffred, T. Heine, An atlas of two-dimensional materials. Chem. Soc. Rev. **43**(18), 6537–6554 (2014). https://doi.org/10.1039/c4cs00102h
58. R.V. Gorbachev, I. Riaz, R.R. Nair, R. Jalil, L. Britnell, B.D. Belle, E.W. Hill, K.S. Novoselov, K. Watanabe, T. Taniguchi, et al., Hunting for monolayer boron nitride: Optical and raman signatures. Small **7**(4), 465–468 (2011). https://doi.org/10.1002/smll.201001628
59. T. Taniguchi, K. Watanabe, Synthesis of high-purity boron nitride single crystals under high pressure by using Ba-BN solvent. J. Cryst. Growth **303**(2), 525–529 (2007). https://doi.org/10.1016/j.jcrysgro.2006.12.061
60. W. Auwärter, T.J. Kreutz, T. Greber, J. Osterwalder, XPD and STM investigation of hexagonal boron nitride on Ni(111). Surf. Sci. **429**(1), 229–236 (1999). https://doi.org/10.1016/S0039-6028(99)00381-7
61. T. Roy, M. Tosun, J.S. Kang, A.B. Sachid, S.B. Desai, M. Hettick, C.C. Hu, A. Javey, Field-effect transistors built from all two-dimensional material components. ACS Nano **8**(6), 6259–6264 (2014). https://doi.org/10.1021/nn501723y
62. S. Beniwal, J. Hooper, D.P. Miller, P.S. Costa, G. Chen, S.Y. Liu, P.A. Dowben, E.C.H. Sykes, E. Zurek, A. Enders, Graphene-like boron-carbon-nitrogen monolayers. ACS Nano **11**(3), 2486–2493 (2017). https://doi.org/10.1021/acsnano.6b08136
63. S. Azevedo, R. De Paiva, Structural stability and electronic properties of carbon-boron nitride compounds. Europhys. Lett. **75**(1), 126–132 (2006). https://doi.org/10.1209/epl/i2006-10066-0
64. S. Fajardo, R.F. García-Galvan, V. Barranco, J.C. Galvan, S.F. Batlle, Graphene-boron nitride composite: A material with advanced functionalities, in *InTech*, ed. by N. Hu, (Rijeka, Croatia, 2012). https://doi.org/10.5772/50729
65. Q. Peng, S. De, Tunable band gaps of mono-layer hexagonal BNC heterostructures. Phys. E Low-Dimensional Syst. Nanostructures **44**(7–8), 1662–1666 (2012). https://doi.org/10.1016/j.physe.2012.04.011
66. J. Zhu, S. Bhandary, B. Sanyal, H. Ottosson, Interpolation of atomically thin hexagonal boron nitride and graphene: Electronic structure and thermodynamic stability in terms of all-carbon conjugated paths and aromatic hexagons. J. Phys. Chem. C **115**(20), 10264–10271 (2011). https://doi.org/10.1021/jp2016616
67. A.Y. Liu, R.M. Wentzcovitch, M.L. Cohen, Atomic arrangement and electronic structure of BC_2N. Phys. Rev. B **39**(3), 1760–1765 (1989). https://doi.org/10.1103/PhysRevB.39.1760
68. Y. Gong, G. Shi, Z. Zhang, W. Zhou, J. Jung, W. Gao, L. Ma, Y. Yang, S. Yang, G. You, et al., Direct chemical conversion of graphene to boron- and nitrogen- and carbon-containing atomic layers. Nat. Commun. **5** (2014). https://doi.org/10.1038/ncomms4193
69. S.Z. Bai, B. Yao, G.Z. Xing, K. Zhang, W.H. Su, Synthesis, conductivity and high-pressure phase transition of amorphous boron carbon nitride. Phys. B Condens. Matter **396**(1–2), 214–219 (2007). https://doi.org/10.1016/j.physb.2007.04.007
70. M.A. Mannan, H. Noguchi, T. Kida, M. Nagano, N. Hirao, Y. Baba, Chemical bonding states and local structures of the oriented hexagonal BCN films synthesized by microwave plasma CVD. Mater. Sci. Semicond. Process. **11**(3), 100–105 (2008). https://doi.org/10.1016/j.mssp.2009.04.003
71. M.A. Mannan, H. Noguchi, T. Kida, M. Nagano, N. Hirao, Y. Baba, Growth and characterization of stoichiometric BCN films on highly oriented pyrolytic graphite by radiofrequency plasma enhanced chemical vapor deposition. Thin Solid Films **518**(15), 4163–4169 (2010). https://doi.org/10.1016/j.tsf.2009.11.086
72. H. Ling, J.D. Wu, J. Sun, W. Shi, Z.F. Ying, F.M. Li, Electron cyclotron resonance plasma-assisted pulsed laser deposition of boron carbon nitride films. Diam. Relat. Mater. **11**(9), 1623–1628 (2002). https://doi.org/10.1016/S0925-9635(02)00047-X

73. R. Zhang, Z. Li, J. Yang, Two-dimensional stoichiometric boron oxides as a versatile platform for electronic structure engineering. J. Phys. Chem. Lett. **8**(18), 4347–4353 (2017). https://doi.org/10.1021/acs.jpclett.7b01721
74. F.M. Arnold, G. Seifert, J. Kunstmann, Thermodynamic stability of borophene, B_2O_3 and other $B_{1\text{-}x}O_x$ sheets. J. Phys. Commun. **4** 031001 (2020). https://doi.org/10.1088/2399-6528/ab7a76
75. C. Zhong, W. Wu, J. He, G. Ding, Y. Liu, D. Li, S.A. Yang, G. Zhang, Two-dimensional honeycomb borophene oxide: Strong anisotropy and nodal loop transformation. Nanoscale **11**, 2468 (2019). https://doi.org/10.1039/C8NR08729F
76. Z. Zhu, X. Cai, C. Niu, C. Wang, Y. Jia, Computational prediction of the diversity of monolayer boron phosphide allotropes. Appl. Phys. Lett. **109**, 153107 (2016). https://doi.org/10.1063/1.4964763
77. P. Ranjan, T.K. Sahu, R. Bhushan, S.S.R.K.C. Yamijala, D.J. Late, P. Kumar, A. Vinu, Freestanding borophene and its hybrids. Adv. Mater. **31**, 1900353 (2019). https://doi.org/10.1002/adma.201900353
78. N.T. Cuong, I. Tateishi, M. Cameau, M. Niibe, M. Umezawa, B. Slater, K. Yubuta, T. Kondo, M. Ogata, S. Okada, I. Matsuda, Topological Dirac nodal loops in non-symmorphic hydrogenated monolayer boron. Phys. Rev. B **101**, 195412 (2020). https://doi.org/10.1103/PhysRevB.101.195412
79. D. Fan, C. Yang, S. Lu, X. Hu, Two-dimensional boron monosulfides: Semiconducting and metallic polymorphs. arXiv 2018, arXiv:1803.03459
80. B. Mortazavi, T. Rabczuk, Boron monochalcogenides; stable and strong two-dimensional wide band-gap semiconductors. Energies **11**, 1573 (2018). https://doi.org/10.3390/en11061573
81. W. Auwärter, H.U. Suter, H. Sachdev, T. Greber, Synthesis of one monolayer hexagonal boron nitride on Ni(111) from B-trichloroborazine $(ClBNH)_3$. Chem. Mater. **16**, 343 (2004). https://doi.org/10.1021/cm034805s
82. J. Osterwalder, W. Auwärter, M. Muntwiler, T. Greber, Growth morphologies and defect structure in hexagonal boron nitride films on Ni(111): A combined STM and XPD study. e-J. Surf. Sci. Nanotech. **1**, 124 (2003). https://doi.org/10.1380/ejssnt.2003.124
83. G.B. Grad, P. Blaha, K. Schwarz, W. Auwärter, T. Greber, Density functional theory investigation of the geometric and spintronic structure of *h*-BN/Ni(111) in view of photoemission and STM experiments. Phys. Rev. B **68**, 085404 (2003). https://doi.org/10.1103/PhysRevB.68.085404
84. W. Auwärter, Hexagonal boron nitride monolayers on metal supports: Versatile templates for atoms, molecules and nanostructures. Surf. Sci. Rep. **74**, 1 (2019). https://doi.org/10.1016/j.surfrep.2018.10.001

Chapter 6
Physical and Chemical Properties of Boron Solids

Ryosei Uno and Kaoru Kimura

6.1 Basic Data about Boron Atom

The boron atom has atomic number 5 and electronic configuration $2s^2 3p^1$ and is a group 13 (IIIb) first-period atom in the periodic table. There are two stable isotopes ^{10}B and ^{11}B, the abundance ratios are 19.9 at% and 80.1 at%, and the atomic weight of the natural element is 10.811. ^{10}B has a large neutron absorption cross section (3840 barns) and is used as a neutron absorber. The first ionization potential of the boron atom is 8.298 eV, the second ionization potential is 25.155 eV, and the covalent radius is 0.88 Å. The melting point is 2077 °C, the boiling point is 3870 °C, and at room temperature boron is usually a black metallic luster solid.

6.2 Crystal Structures

α-rhombohedral, β-rhombohedral, α-tetragonal, β-tetragonal, and γ-orthorhombic boron have been reported as solid boron crystals with B_{12} icosahedron. Boron solids are usually made by bringing hydrogen containing gaseous boron halide into contact with a hot tantalum wire and depositing the hydrogen reduced boron on the tantalum wire surface. When the temperature of the tantalum wire is 900 to 1000 °C, α-rhombohedral boron is deposited with amorphous boron through the TaB_2 layer; at 1200 °C or more, β-rhombohedral boron are deposited; and at 1150

R. Uno
Nihon University, Tokyo, Japan

K. Kimura (✉)
The University of Tokyo, Tokyo, Japan
e-mail: bkimura@phys.mm.t.u-tokyo.ac.jp

I. Matsuda, K. Wu (eds.), *2D Boron: Boraphene, Borophene, Boronene*,
https://doi.org/10.1007/978-3-030-49999-0_6

Fig. 6.1 Temperature-pressure phase diagram of pure boron [3]

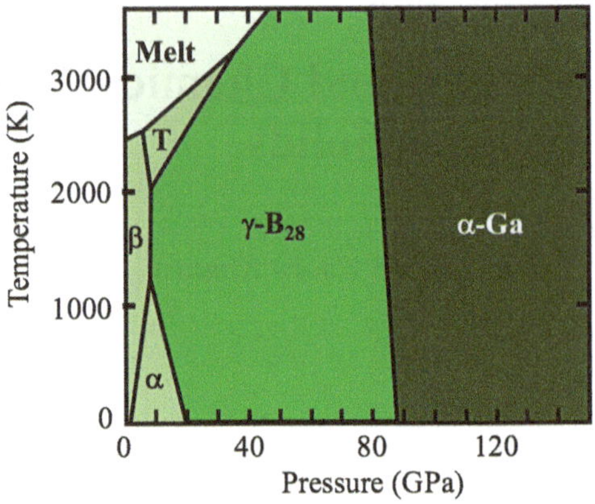

to 1200 °C, β-tetragonal boron are deposited in a narrow range [1, 2]. Recently, the temperature-pressure phase diagram of pure boron has been constructed [3, 4] as shown in Fig. 6.1. At ambient pressure, β-rhombohedral boron is stable in all temperature range. α-rhombohedral and β-tetragonal boron are stable at high pressure and low- and high-temperature range, respectively. At higher pressure, γ-orthorhombic boron is stable, and then α-Ga-type crystal without B_{12} icosahedron becomes stable. Although α-tetragonal boron was confirmed to be $B_{50}C_2$ or $B_{50}N_2$ instead of pure boron, recently, it's surface energy has been found to be lower than those of the other crystals [5], and it was realized as the form of nanostructure such as nano-belt [6].

The α-rhombohedral boron is made of only B_{12} icosahedra, and the β-rhombohedral and β-tetragonal boron each include a B_{28} structure and a B_{21} structure, in which a plurality of B_{12} icosahedra is merged. In addition, it has a B single atom in the void in a skeletal structure formed from B_{12} and B_{28} or B_{21}. There are reports of cubic crystals produced by plasma discharge [7].

6.2.1 α-rhombohedral Boron

The crystal structure of α-rhombohedral boron is shown in Fig. 6.2(a) and (b). Figure 6.2(b) is a projection of the rhombohedral crystal onto the (111) plane [hexagonal (0001) plane]. There is one B_{12} regular icosahedron at the vertex of the rhombohedron, and the threefold symmetry axis of B_{12} faces the rhombohedral body diagonal direction. This B_{12} is slightly distorted and connected by a B-B bond between the B atoms at the vertex, and the direction of the B-B bond between B_{12} is slightly shifted from the direction of the edge of the rhombohedron. The rhombohedral crystal has a lattice constant of a = 5.074 Å and α = 58.0 deg., and the unit cell contains twelve B atoms [8]. The B atom that connects B_{12} with the two-center covalent bond is called a polar atom (p), and the other B atoms are arranged

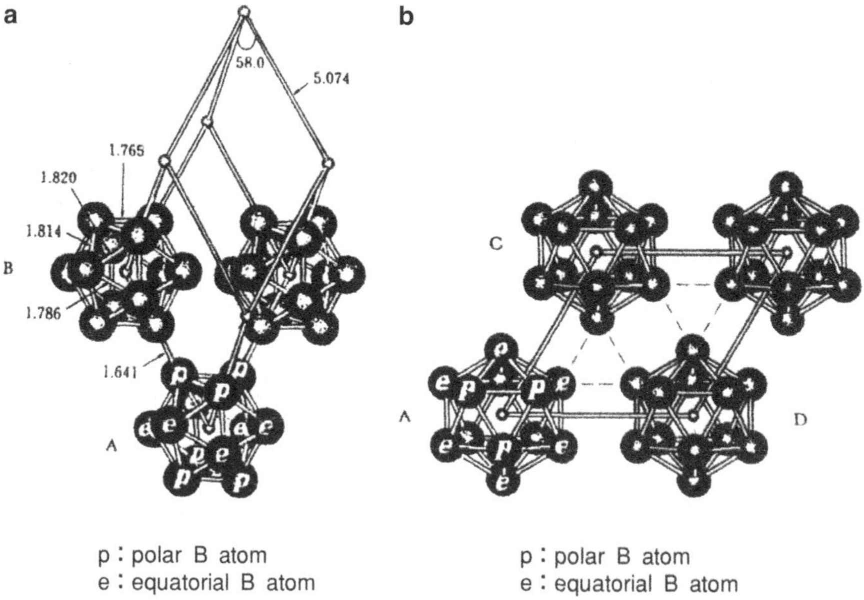

Fig. 6.2 (**a**, **b**) Crystal structure of α-rhombohedral boron. At each vertex of the rhombohedron, there is a regular icosahedron made of B atoms, and its threefold axis is parallel to the rhombohedral body diagonal. In (**a**), three B_{12} icosahedra are shown at the vertices, and the others are omitted to make the figure easier to understand. The six atoms involved in the bonds along the rhombohedral edges between B_{12} are called polar atoms, and the other six atoms are called equatorial atoms. B_{12} is distorted and the interatomic distance is shown in Å. (**b**) shows the (111) rhombohedral coordinate plane. It corresponds to the bottom plane in hexagonal coordinates. (Reprinted by permission from R. Uno, K. Kimura, Chap. 1, Fundamentals and Applications of Boron, Borides and Related Materials (CMC Publishing Co., LTD. 2008). Copyright (2008) CMC Publishing Co.)

in a zigzag regular hexagon in the plane perpendicular to the threefold symmetry axis and are called equatorial atoms (e). The average length of the B-B bond in B_{12} is 1.79 Å, but the length of the B-B bond between B_{12} is 1.64 Å, so it is difficult to think of it simply as a B_{12} icosahedron bond. This crystal also has a large void in the center of the rhombohedron.

6.2.2 *β-Rhombohedral Boron*

β-rhombohedral boron has a complex structure with 20 types of B atom sites in the rhombohedral unit cell. The lattice constant of the rhombohedral crystal is a = 10.143 Å, α = 65.22 deg. [9]. The basic structure is formed of 15 types of atomic sites B1 to B15 shown in Table 6.1 [9] and has 105 B atomic sites per unit cell (315 for hexagonal unit cell). The structure is made up of a B_{12} icosahedron, a structure

Table 6.1 Position coordinates of B atom in β-rhombohedral boron [9]

B atom site	Wyckoff position	Hexagonal coordinates			Occupancy %
		x	y	z	
B1	36i	0.1730	0.1741	0.1767	100
B2	36i	0.3191	0.2957	0.1294	100
B3	36i	0.2614	0.2175	0.4197	100
B4	36i	0.2349	0.2516	0.3469	100
B5	18 h	0.0545	2 x	0.9437	100
B6	18 h	0.0865	2 x	0.0132	100
B7	18 h	0.1098	2 x	0.8860	100
B8	18 h	0.1705	2 x	0.0280	100
B9	18 h	0.1287	2 x	0.7666	100
B10	18 h	0.1020	2 x	0.6985	100
B11	18 h	0.0564	2 x	0.3266	100
B12	18 h	0.0897	2 x	0.3990	100
B13	18 h	0.0579	2 x	0.5538	74.5
B14	6c	0.0	0.0	0.3855	100
B15	3b	0.0	0.0	0.5000	100
B16	18 h	0.0546	2 x	0.1176	27.2
B17	18 h	0.0833	2 x	0.4760	8.5
B18	18 h	0.1440	2 x	0.5239	6.6
B19	18 h	0.1805	2 x	0.5347	6.8
B20	36i	0.2067	0.2280	0.0711	3.7

where half of the B_6 pentagonal pyramids are merged, and an isolated B single atom. There are two ways to understand the structure. The first is centered on B_{12}, with a pentagonal pyramid connected to each vertex, with a B_{84} structure similar to a soccer ball at the vertex of the rhombohedron, and with a B_{10} structure merged with three pentagons and an aggregate with a B_{10}-B-B_{10} chain structure consisting of B single atoms in the center of the rhombohedron. The other is an assembly of four B_{12}, three B_{12} merged B_{28} structures, and a B_{28}-B-B_{28} chain structure consisting of B single atoms.

The latter structure is shown in Fig. 6.3(a). The figure also shows the main voids of this structure. Figure 6.3(b) shows a B_{28} structure composed of three B_{12} sharing three surfaces. Figure 6.3(c) shows the bonds between the B atoms that represent the connections between the structures. The positions of the B atom sites are classified into 15 types based on symmetry, and the positions indicated by hexagonal coordinates are shown in Table 6.1. Of these, the B13 site is in a state of stable deficiency, and most of the deficient atoms occupy B16 of the B16 to B20 sites that correspond to the voids in the basic structure. The B13 and B16 site occupancy rates are 0.745 and 0.272, respectively. 105.1 B atoms occupy the B1 to B16 sites. There are a few other B atoms that occupy B17 to B20 sites, and when they are added, there are 106.9 B atoms per rhombohedral unit cell.

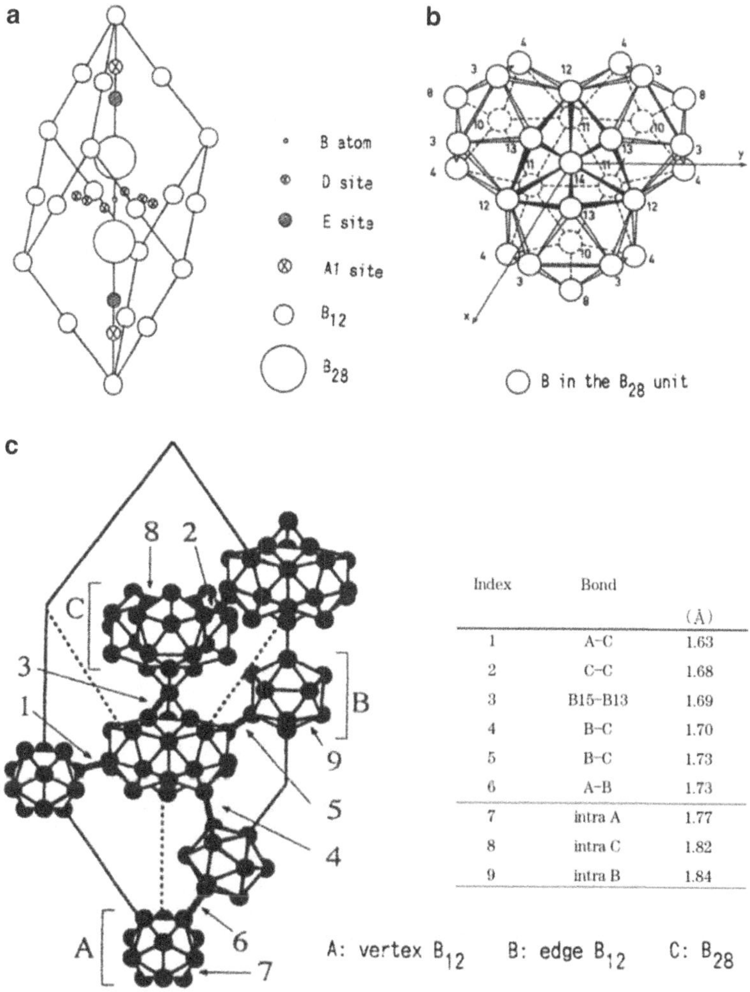

Index	Bond	(Å)
1	A–C	1.63
2	C–C	1.68
3	B15–B13	1.69
4	B–C	1.70
5	B–C	1.73
6	A–B	1.73
7	intra A	1.77
8	intra C	1.82
9	intra B	1.84

Fig. 6.3 (**a**, **b**, **c**) β-rhombohedral boron. Since the unit cell has 105 atomic sites and has a complex structure, (**a**) shows a conceptual diagram with a relatively large void, A1 site, D site, and E site. (**b**) shows the B_{28} structure as seen from the body diagonal direction. The numbers indicate the B atom site numbers. (**c**) shows the connection between B atoms and the interatomic distance [9] 7, 8, and 9 are average values. Unnecessary B_{12}s are omitted to make the figure easier to understand. The single atom at the center of the rhombohedron between B_{28}s is the B15 atom. (Reprinted by permission from R. Uno, K. Kimura, Chap. 1, Fundamentals and Applications of Boron, Borides and Related Materials (CMC Publishing Co., LTD. 2008). Copyright (2008) CMC Publishing Co.)

6.2.3 α-Tetragonal Boron

Although it is not a pure boron structure, it is closely related to the boride structure. The structure is shown in Fig. 6.4. The lattice constants are a = 8.75 Å, c = 5.06 Å, and there are four B_{12} in the unit cell and two B single atoms in total at the vertex and body center of the unit cell, for a total of 50 B atoms. Furthermore, it has a total

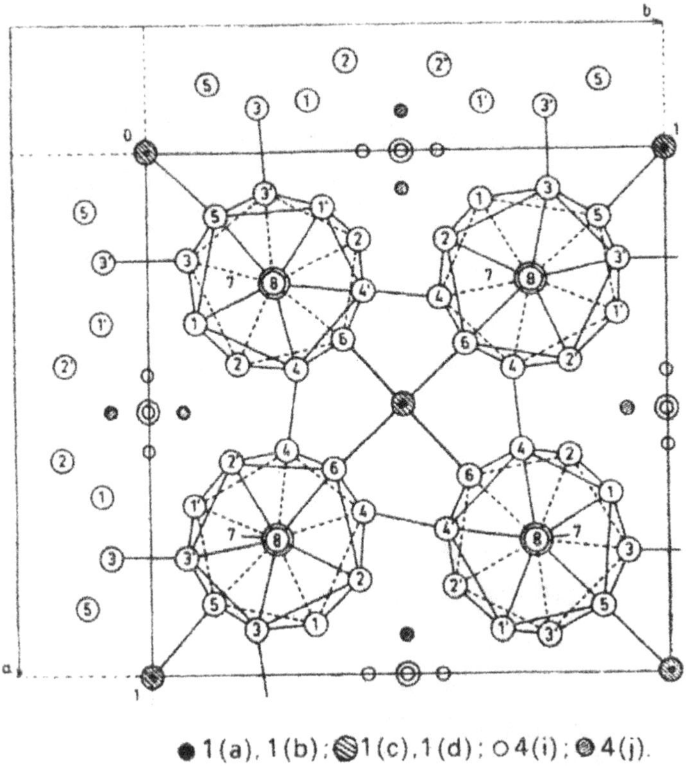

Fig. 6.4 Crystal structure of α-tetragonal boron. A projection view on the (001) plane. 1(a) and 1(b) are the positions of the body center, and 1(c) and 1(d) are the positions shifted by c/2 in the c-axis direction, occupied by C atoms or N atoms. The center of 4(i), 4(j), and 4(i) is a void where different atoms can easily enter. (Reprinted by permission from R. Uno, K. Kimura, Chap. 1, Fundamentals and Applications of Boron, Borides and Related Materials (CMC Publishing Co., LTD. 2008). Copyright (2008) CMC Publishing Co.)

of two C single atoms or N single atoms in the center of the c-axis direction, where B single atoms are arranged, and there is a total of 52 atoms in the unit cell [10].

Recently discovered nano-belts of pure boron have this crystal structure (50 per unit cell) [6] because the surface energy of this structure is lower than others [5].

6.2.4 *β-Tetragonal Boron*

It has a complex structure consisting of a two-center or three-center covalent bond, in which two or three B atoms share two electrons each. The lattice constants are a = 10.14 Å and c = 14.17 Å, and the unit cell contains 190 B atoms. A projection of a structure in which four chain structures of B_{12} icosahedra are arranged, penetrate the unit cell perpendicular to the c-axis, and increase in height by 0.25c perpendicular to each other is shown in Fig. 6.5(a), together with a projection

a

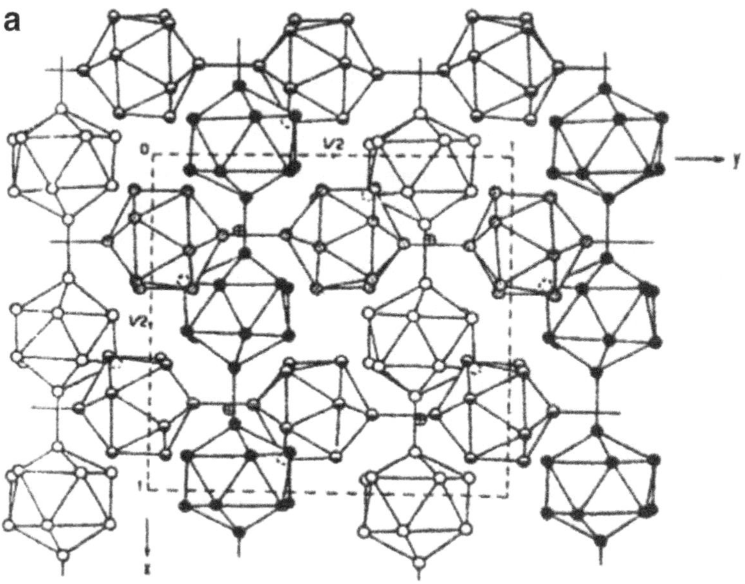

○B in B_{12} (z=0.10) ◉B in B_{12} (z=0.35) ●B in B_{12} (z=0.60)
◉B in B_{12} (z=0.85) ⊕interstitial B26

b

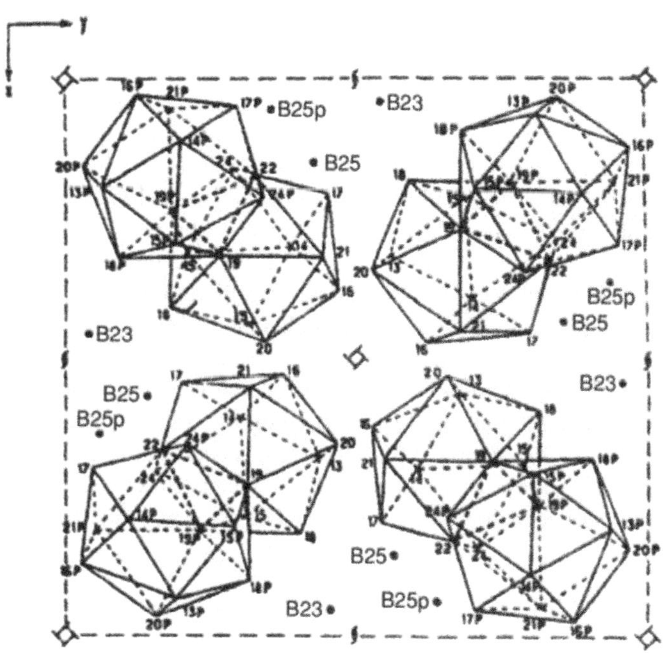

Fig. 6.5 (**a**, **b**) Crystal structure of β-tetragonal boron. It has a very complex structure with 190 atoms in the unit cell. Both (**a**) and (**b**) are projections onto the (001) plane, each showing part of the structure. (**a**) is a projection of a cluster atom chain connected by B_{12}s and two single atoms (B26 filling rate 0.5), with four chains penetrating the unit cell. (**b**) is a projection of four B_{21}

of two single atoms per unit cell on a plane perpendicular to the c-axis (four sites with an occupancy of 0.5) [11]. There is a total of eight B_{12} icosahedra, two for each chain per unit cell. In the opening of this chain structure, the B_{21} structure, in which two B_{12}s are joined together by sharing a triangular surface, places its center in the void near (1/4, 1/4, 1/4), (3/4, 3/4, 3/4), (1/4, 3/4, 1/2), and (3/4, 1/4, 0). The projection onto the (001) plane is shown in Fig. 6.5(b) together with the projection of eight single atoms per unit cell. Eight of the twelve single atom sites are said to be in a stable deficient state. The occupancy ratio is 0.5 for (a) B26 in the Figure, and (b) B25 and B25p in the Figure, respectively. B23 has an occupancy ratio of 1.

6.2.5 γ-Orthorhombic Boron

The phase is stable under the pressure of 19–89 GPa, and it can be quenched to ambient condition [3]. The space group is *Pnnm*, and there are 28 atoms in the unit cell, consisting of icosahedral B_{12} clusters and B_2 dumbbells in a NaCl-type arrangement, as shown in Fig. 6.6(a) [83]. The structure was determined by X-ray diffraction experiment. Because of the charge transfer between these two cluster components, γ-B_{28} can be regarded as a boron boride, $(B_2)^{\delta+}(B_{12})^{\delta-}$. The phase indicates ionic high-pressure form of elemental boron [3].

6.3 Characteristics of Boron Icosahedral Clusters

Boron crystals are based on the B_{12} icosahedron. The icosahedron is one of the five regular polyhedrons in the convex polyhedrons of Euclidean space: regular tetrahedron, regular hexahedron (cube), regular octahedron, regular dodecahedron, and regular icosahedron. As shown in Fig. 6.7, the regular icosahedron has five, three, and two axes of symmetry, but the fivefold symmetry is incompatible with the translational symmetry of the crystal. It is therefore a problem to fill the space, and the icosahedron is distorted as a whole in the crystalline structures described in the previous section. The icosahedral B_{12} cluster has a structure with two missing valence electrons, as will be described later, and is distorted due to the Jahn-Teller effect. It is thought that the symmetry is lost, the orbital degeneracy is solved, and the empty orbital is separated from the orbitals occupied with electrons.

Figure 6.8 shows the experimentally obtained electron density distribution in the α-rhombohedral boron, in which two-center and three-center covalent bonds

Fig. 6.5 (continued) structures and eight single atoms (B25, B25p filling rate 0.5) in a four-chain void. B_{21} is a paired structure of B_{12} sharing one surface. (Reprinted by permission from R. Uno, K. Kimura, Chap. 1, Fundamentals and Applications of Boron, Borides and Related Materials (CMC Publishing Co., LTD. 2008). Copyright (2008) CMC Publishing Co.)

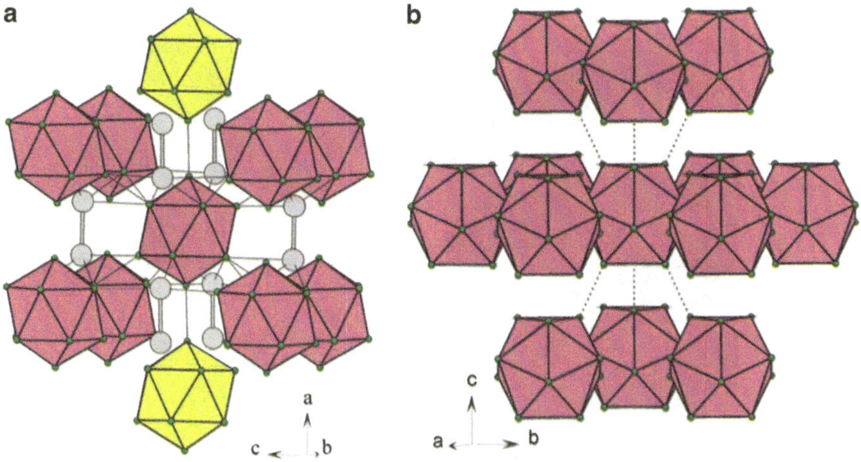

Fig. 6.6 Atomic structures of (**a**) γ -B_{28} [83]. A crystal of (b) α -B_{12} is shown for comparisons. The ionic γ -B_{28} structure is composed of the two cluster components of B_{12} icosahedra and B_2 dumbbells with the opposite charges, as a boron boride, $(B_2)^{\delta +}$ $(B_{12})^{\delta -}$. (Reprinted by permission from U. Häussermann and A. S. Mikhaylushkin, Structure and Bonding of γ -B_{28}: Is the High Pressure Form of Elemental Boron Ionic? Inorg. Chem. 49, 11,270 (2010). Copyright (2010) American Chemical Society)

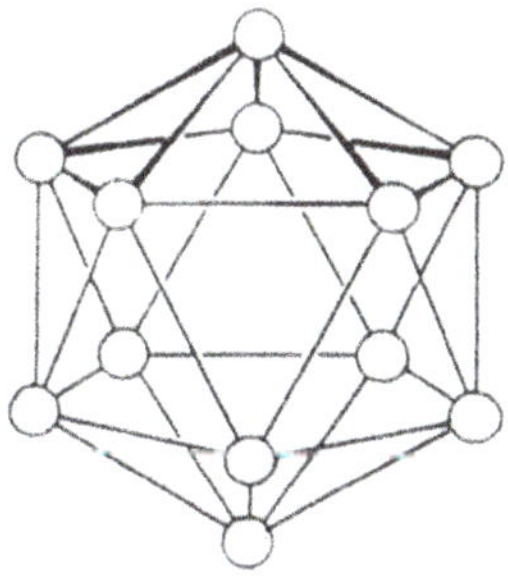

Fig. 6.7 B_{12} icosahedron In the projection from the threefold axis direction, the axis connecting the vertices of the pentagonal pyramids facing each other is the fivefold axis. The axis connecting the midpoints between the B atoms on the left and right sides of the Figure is the twofold axis. (Reprinted by permission from R . Uno, K. Kimura, Chap. 1, Fundamentals and Applications of Boron, Borides and Related Materials (CMC Publishing Co., LTD . 2008). Copyright (2008) CMC Publishing Co)

are clearly observed [12]. In the α-rhombohedral boron made of B_{12} alone, in the regular triangle surface formed by B_p atoms, the electron density is higher in the center of the triangle than in the center of the lines connecting the atoms as shown in Fig. 6.8(c), and B_p is joined with a three-center covalent bond. The triangular faces formed by the B_p, B_p, B_e atoms and the B_p, B_e, B_e atoms are also joined by a similar three-center covalent bond. Also, between B_{12}, B_e, B_e, B_e atoms shown in Fig. 6.8(b) are joined by a weak three-center covalent bond.

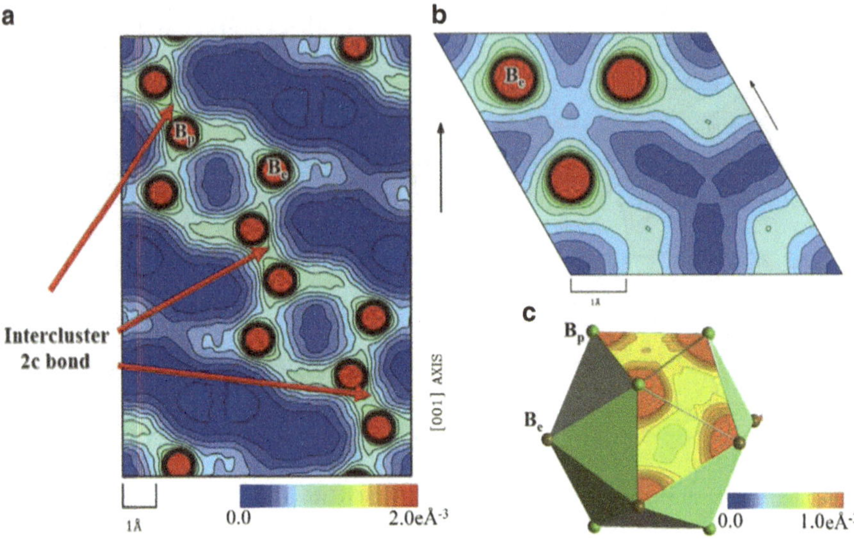

Fig. 6.8 MEM charge densities of α-rhombohedral boron [12]. (**a**) The MEM charge density for 110 plane. The B_P-B_P 2c intercluster bond is located on this plane. (**b**) The MEM charge map for 001 plane. The 3c intercluster bond is located on this plane. (**c**) The MEM charge density of intracluster bonds. (Reproduced from Ref. [12])

The three-center covalent bond is unique to group III elements. A normal two-center covalent bond appears to double the number of electrons due to two atoms sharing each other's electrons. As shown in Fig. 6.9 [13] (a)–(c), the atoms of groups IV, V, and VI appear to have eight valence electrons by making four, three, or two such bonds. The s and p orbitals are occupied completely by the eight electrons (closed shell structure). However, as shown in Fig. 6.9(d), group III atoms have only six valence electrons even if they form three covalent bonds, and the bonds cannot be completed. Therefore, as shown in Fig. 6.9(e), two valence electrons are used for the three-center covalent bonds that share electrons with three atoms, appearing to triple the number of electrons, and one valence electron is used for the usual 2-center covalent bond. By saving electrons in this way, the atom can appear to have eight valence electrons. A regular icosahedron cluster is composed of regular triangular faces. It is a convenient structure to realize the above situation by using two valence electrons for bonds inside the cluster and one valence electron for bonds outside the cluster. Therefore, the icosahedral cluster of twelve atoms is suitable for group III elements to form covalent bonds.

On the other hand, an icosahedral cluster of 13 atoms with an atom at the center of the cluster has a packing fraction close to that of a close-packed structure and is a structure suitable for metallic bonding. For these reasons, the icosahedral cluster of boron and aluminum at the boundary between the element that forms covalent bonds and the element that forms metallic bonds in the periodic table causes a metallic-covalent bonding conversion due to the presence of the central atom [14, 15].

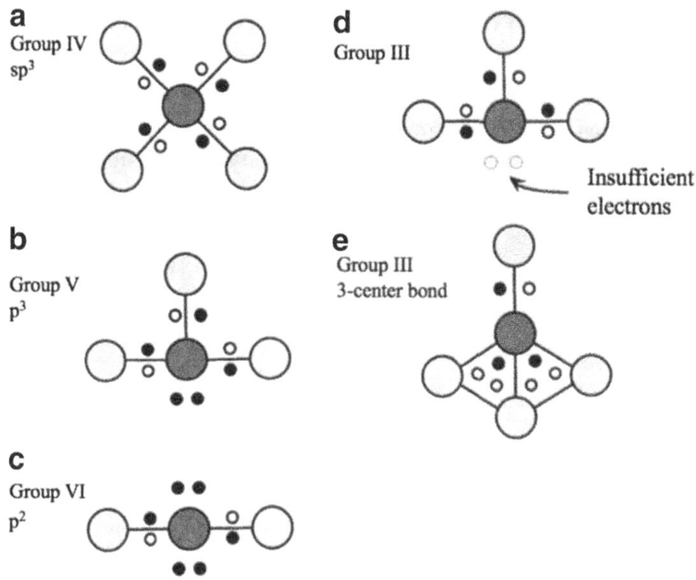

Fig. 6.9 Schematic diagram of two-center and three-center covalent bonds [13]. (Reproduced and translated from Ref [13])

More precisely, in the molecular orbital calculation for an isolated B_{12} icosahedron cluster, there are twelve orbitals in the fivefold direction used for bonding to the outside of the cluster, and 13 orbitals inside the cluster, so that 12 + 13 x 2 = 38 electrons are needed to satisfy the orbitals, but there are only 12 x 3 = 36 valence electrons, and there are still two electrons deficient [16]. This induces the fact that the band structure of a boron crystal has an acceptor level with a high density. In the calculation result by the self-consistent molecular orbital method for the $B_{12}H_{12}$ cluster, which approximates the state in the crystal by bonding H atoms to the outside of each B atom of B_{12}, the highest level, which is degenerate fourfold when there is no distortion, becomes no longer degenerate due to the Jahn-Teller effect and the symmetry changes from Ih to D3d or D2h [17]. In the D3d symmetry corresponding to the β-rhombohedral boron, the icosahedron that is stabilized by molecular orbital calculation has the same distortion as B_{12} of the β-rhombohedral boron. At this time, the level of fourfold degeneracy is divided into a twofold degenerate level and a single level complete with electrons and an empty single level. Therefore, in the rhombohedral crystal, the orbital of the highest level of the valence band is separated due to the Jahn-Teller effect, and the intrinsic acceptor level with a large density of states is about 0.2 eV above the full valence band [18].

In the α-rhombohedral boron, electrons can be saved by the three-center bond between clusters, so the number of electrons used for bonding outside the cluster is reduced to ten and the orbitals in the cluster are completely filled. For this reason, there is no intrinsic acceptor level.

6.4 Physical Properties of β-Rhombohedral Boron

The boron solids that are usually made are β-rhombohedral boron. Room-temperature β-rhombohedral boron is chemically inert, light, and hard and has a large elastic constant and a high sound velocity. Boron fibers are expensive but have high performance as composite reinforcements and are used in aircraft materials.

6.4.1 Basic Properties

The density is 2.34 x 10^3 kg/m^3, the Mohs hardness is 9.3, the Vickers hardness is 49,000 MPa, the bulk modulus is 185 GPa, and the sound velocity reaches 16,200 m/s. The thermal conductivity at room temperature is 27.4 W/m·K, and the coefficient of linear thermal expansion is 5 to 7 x 10^{-6} /K.

6.4.2 Doping Sites (Crystal Structure Voids)

If the covalent bond radius $r_B = 0.88$ Å is assumed for the size of the boron atom, this structure has many voids because the packing fraction of a β-rhombohedral boron is only 0.36. The main positions are shown in Table 6.2 [19, 20]. These voids first attracted attention because, in the structural analysis of sample $CrB_{\sim 41}$, which investigate the solid solubility of Cr, the position of B atoms was almost the same as β-rhombohedral, and the Cr atoms were statistically distributed in the two voids A1 and D [21]. Later, many metal atom solid solutions were investigated, and it was found that the positions of the voids distributed differed depending on the type of atom, and even when the same voids were occupied, the positions shifted slightly. The main results are shown in Table 6.3, and the main positions are shown in Fig. 6.10(a) and (b). The atom occupying A3 has not been found. The different atoms occupying the voids supply electrons to the band of the β-rhombohedral boron and change the electrical properties. V atoms are not strongly doped, and the semiconductor-metal transition occur without the crystal structure changing significantly. In general, when the number of dissolved atoms increases, valence electrons are supplied to the electron-deficient B_{12} structure, and the complex skeleton structure of the B atoms changes.

6.4.3 Electrical Properties

6.4.3.1 Electrical Conductivity

The β-rhombohedral boron is a semiconductor, and the electrical conductivity of a pure sample at room temperature is 10^{-6} to 10^{-7} Ω^{-1} cm^{-1} and the activation

Table 6.2 Doping sites (voids) of β-rhombohedral boron [19, 20]. G is the unoccupied remaining void of B16 (occupancy approximately 25%). The position coordinates change slightly depending on the type of atom being doped

Void site	Wyckoff position	Hexagonal coordinates	Number of adjacent B atoms Distance (Å)
A1	6c	0.0; 0.0; 0.135	12
			2.15–2.19
A2	18 h	0.108; 0.216; 0.100	12
			2.06–2.14
A3	36i	0.279; 0.248; 0.046	12
			1.91–2.20
D	18 h	0.205; 0.411; 0.174	1; 14
			2.01;
			2.28–2.45
E	6c	0.0; 0.0; 0.245	15
			2.23–2.53
F1	18 h	0.065; 0.130; 0.245	7; 2
			1.95–2.18;
			2.79
F2	18 h	0.117; 0.233; 0.252	3; 13
			2.12;
			2.39–2.66
G	18 h	0.055; 0.109; 0.117	8; 6
			1.79–1.85;
			2.54–2.62
H	18 h	0.208; 0.792; 0.412	11; 1
			2.19;
			2.29

energy is 1.56 eV. The carriers are holes and exhibit p-type electrical conduction. In silicon, which is a typical semiconductor, doped impurity atoms replace silicon atoms to create acceptor levels and donor levels. But in β-rhombohedral boron, as suggested by the B_{12} icosahedron structure, the acceptor level concentration is large even in a pure solid, and p-type conduction is exhibited for most of the doped impurities at low concentrations. However, even samples doped with relatively high concentrations of Fe or low concentrations of V, etc. exhibit n-type conduction [26–28].

Figure 6.11 [29] shows the temperature dependence of electrical conductivity for high-purity β-rhombohedral boron together with samples containing various impurities. The activation energy of a pure sample is 1.56 eV. Figure 6.12 (a) [30] and (b) [27] show the temperature dependences of various samples below room temperature. This result shows that the electrical conduction follows the following formula, which shows Mott variable range hopping conduction.

$$\sigma(T) = \sigma_0 \exp\left\{-(T_0/T)^{1/4}\right\}, T_0 = 60\alpha^3/\pi N(E_F)k_B$$

Table 6.3 Doping sites (voids) and occupancy of solid solution atoms in solid solutions of transition metal atoms and atoms with p-electron deficiencies in β-rhombohedral boron

		(β-B)	Li	Mg	Al	Si	Sc	V	Cr	Mn	Fe	Ni	Cu	Ge	Zr
Solid solution atomic site	Wyckoff position	Atom Occupancy (%)													
A1	6c				Al 84.1	Si 46.4		V 64.0	Cr 71.9	Mn 25.6	Fe 50.7	Ni 44.7	Cu 7.9	Ge 20.9	
A2	18 h					Si 4.8								Ge 0.46	
D	18 h		Li 100.0	Mg 60.2	Al 21.6		Sc 31.4	V 5.3	Cr 18.0	Mn 43.1	Fe 18.5	Ni 21.6	Cu 33.4	Ge 2.9	Zr 27.9
E	6c		Li 100.0	Mg 86.1			Sc 72.7			Mn 66.2			Cu 59.9		Zr 18.1
F	18 h			Mg 21.6											
H	18 h			Mg 19.7											
B13	18 h	B 74.5	B 64	B 69.8		B 73.9	B 61		B 72	B 65	B 73	B 67.1	B 69	B 71.6	B 53
B16	18 h	B 27.2	B 10									B 18.0	B 13		B 14.5
Replacement position	36i			B4		B1	B4							B1	
Atom occupancy (%)				B 83.0		Si 13.3	Sc 6.0							Ge 3.4	
References		[9]	[14, 23]	[20, 25]	[22]	[19]	[19]	[14]	[19]	[19]	[19, 24]	[14, 19]	[19, 22]	[19]	

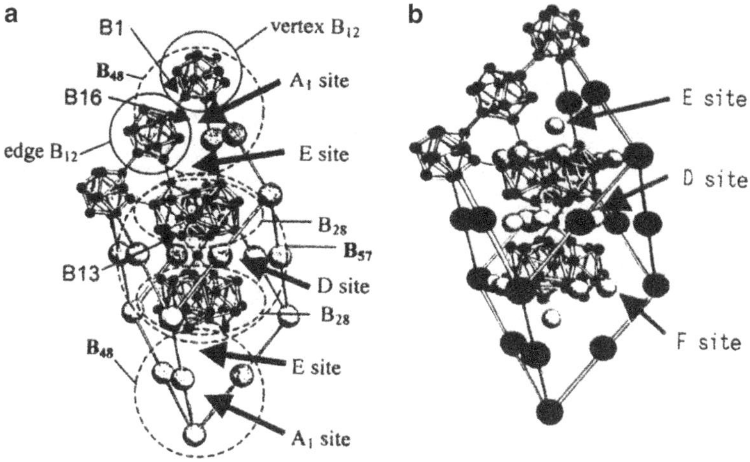

Fig. 6.10 (**a**, **b**) Main voids in β-rhombohedral boron. The A1, D, and E sites are shown in (**a**), and the D, E, and F sites are shown in (**b**). Also, (**a**) shows that the β-rhombohedral boron can be seen to be composed of a B_{48} structure consisting of four B_{12}s at the vertex of the rhombohedron and a B_{57} structure consisting of B_{28}-B15-B_{28} contained in the rhombohedron. (Reprinted by permission from R. Uno, K. Kimura, Chap. 1, Fundamentals and Applications of Boron, Borides and Related Materials (CMC Publishing Co., LTD. 2008). Copyright (2008) CMC Publishing Co.)

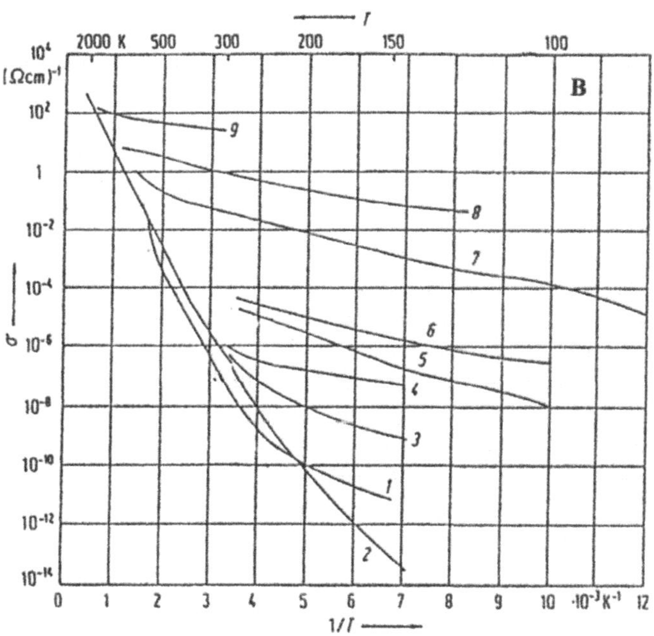

Fig. 6.11 Temperature dependence of electrical conductivity of β-rhombohedral boron[29)0].1, 2: High-purity boron. 3: Surface polishing. 4: B_6O surface layer. 5: Mn dopant. 6: Fe dopant. 7: 3% C dopant. 8: $FeB_{29.5}$. 9: C dopant ($p = 3 \times 10^{20}/cm^3$). (By courtesy of H. Werheit)

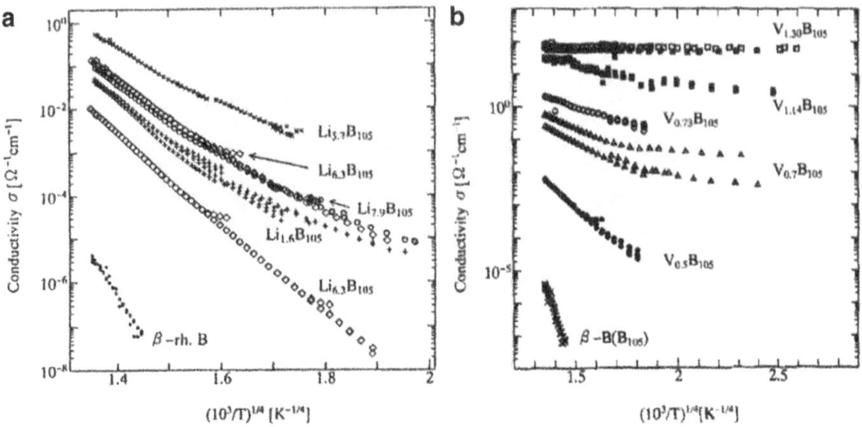

Fig. 6.12 (**a**) [30], (**b**) [27] Temperature dependences of electrical conductivity when various amounts of Li (in (a)) or V (in (**b**)) are doped into β-rhombohedral boron. (Reproduced from Ref. [30, 27])

Here, k_B is the Boltzmann constant, $N(E_F)$ is the density of states at the Fermi level, and α is the damping constant (the inverse of localization length) of the wave function in that state. The fact that p-type conduction is hopping conduction shows that the holes are localized due to large disorder (defects and impurities) and electron-lattice interactions. This is despite the relatively large density of states at the top of the valence band due to B_{12}.

It also suggests that hopping conduction and band conduction coexist near 150 °C, and even if the Hall effect is observed, the contribution of carriers that perform band conduction is large, so problems remain in the carrier concentration and mobility obtained from the Hall constant calculated by the current including the contribution due to hopping conduction [31].

When the DC applied voltage at room temperature reaches about 10^4 V/cm, the resistance suddenly becomes low, and the so-called switching phenomenon is observed in which a high resistance state and a low resistance state are repeated with a recovery time of about 10 μs. However, in the measurement using a single pulse voltage with a width of 10 μs, no switching phenomenon is observed up to 60 kV/cm even at a low temperature of 98 K, which is considered to be due to the temporarily trapped carriers at the hopping sites [32].

Doped atoms rarely enter interstitial doping sites to replace boron atoms, and their valence electrons are considered to enter the acceptor level due to B_{12} clusters and defects. However, the site to enter depends on the type of atom, and not only the size but also the contribution to the bond with the B atom is assumed to be different. It is considered that metallic-covalent bonding conversion occurs in the doped samples of V atoms [14], and the semiconductor conduction approaches metallic conduction with about 1 at% doping.

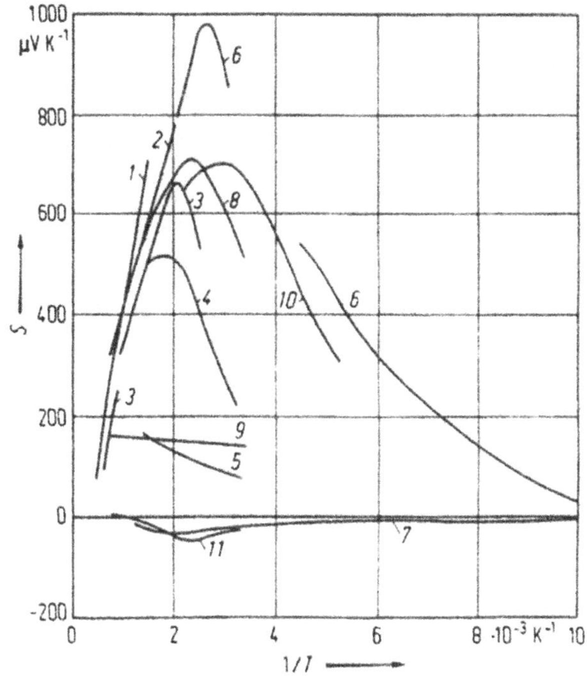

Fig. 6.13 Temperature dependence of the Seebeck coefficient of β-rhombohedral boron [29]. The measurement data from various samples is shown. 1. Sh. Z. Dzhamagidze et al., 2. O. A. Golikova et al., 3. H. Werheit et al., 4. Industrial samples J. Jaumann et al., 5. R. Uno 6. A. Zareba et al., 7. Fe doped, 8. C doped A. A. Berezin et al., 9. C doped A. A. Berezin et al., 10. G. Majni et al., 11. Fe doped J. M. Dusseau et al. (By courtesy of H. Werheit)

When doping with Li, Cu, Ni, Mg, etc., the amount varies depending on the element, but the conductivity becomes maximum when the doping amount is 2 to 5 per unit cell, and if it is doped more, it has been observed that the conductivity may decrease instead [27, 30]. In the range of the doping amount where semiconductor conduction is observed, variable region hopping conduction is observed for all elements as shown in Fig. 6.12(a) and (b).

The thermoelectric power data varies widely. The Seebeck coefficient is positive (p-type conductivity) for most samples, and the samples that seem to have a high C concentration have a higher coefficient at higher temperatures, such as boron carbide. High-purity samples, etc. decrease from 400 to 500 K at a peak of about 700 μV/K at higher temperatures [29]. The accumulated data is shown in Fig. 6.13. Samples doped with Fe, V, etc. exhibit negative values. A unique feature is that it shows about 200 μV/K even at high temperatures exceeding 1000 K. As a thermoelectric material, the p-type is not as effective as boron carbide that maintains 300 μV/K without reducing the Seebeck coefficient even if it exceeds 1500 K, but samples doped with V are n-type and have a performance close to boron carbide [33].

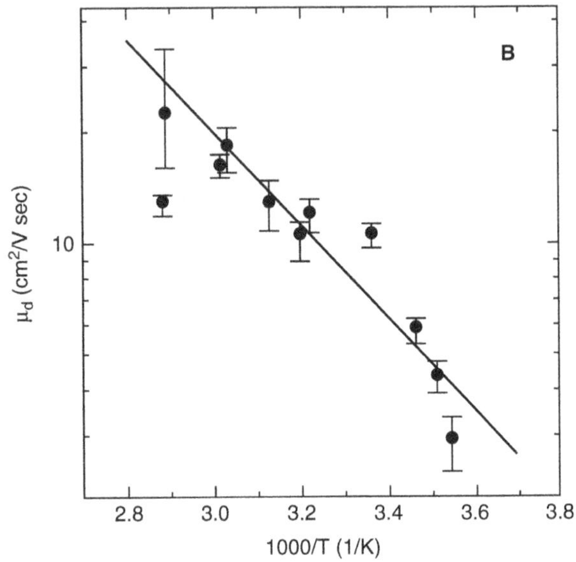

Fig. 6.14 Temperature dependence of drift mobility for holes in β-rhombohedral boron measured by transient photoconduction [35]. The activation energy is 0.24 eV. (Reproduced from Ref. [35])

Regarding the Hall effect, it can be said that there is no reliable data for high-purity samples because the effect at room temperature is small. However, positive Hall effect showing p-type conduction at a high temperature of 600 K has been reported [34]. Since the Hall effect does not occur in hopping conduction, it is considered that a current occurs due to band conduction at high temperatures. When determining the Hall coefficient, it is necessary to estimate the current due to band conduction alone.

In electrical conduction, carrier mobility is the most interesting physical quantity, but very different values are reported depending on the phenomenon used in the evaluation, and the values at room temperature range from 10^{-5} to 3 x 10^{+2} cm^2/V·s [29]. Since there is a high possibility that hopping conduction and band conduction coexist, future research is required to evaluate mobility. As for the value of drift mobility, the most reliable measurement value can be obtained if an elementary process can be measured with a very short pulse of light in transient photoconduction. The reported value is about 8 x 10^{-2} cm^2/V·s at 300 K for holes. Figure 6.14 shows the temperature dependence of the drift mobility for holes obtained from transient photoconduction. The drift mobility has an activation energy of 240 meV, indicating that it is not normal band conduction [35].

Furthermore, in measurements below room temperature, the space charge produced by carriers trapped at the trap level is considered to be steady state or take a long time to disappear after measurement, and there is a hysteresis phenomenon that seems to be originated from the space charge. For this reason, there is often a problem in the reproducibility of data, and evaluation of data is often difficult. In the case of industrial use, it can be processed with data according to the practical state, but it is difficult to process the data for academic use.

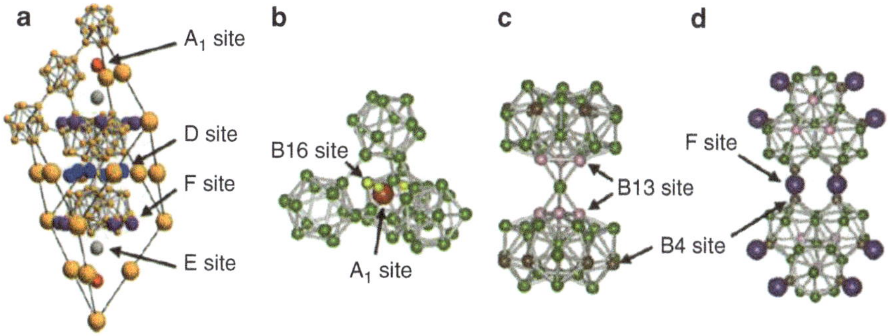

Fig. 6.15 Structure and several sites of the β-rhombohedral boron, B_{105} [37]. (**a**) The major doping sites or crystal structure voids are indicated. The small, medium, and large spheres correspond sites for single B atoms, dopants, and B_{12} clusters, respectively. Recalling Fig. 6.10, structures of (**b**) B_{48} cluster, (**c, d**) B_{57} cluster are shown with the site labels. In actual the β-rhombohedral boron, $B_{106.6}$, the occupancies of B13 and B16 sites are approximately 75% and 25%, respectively, while those of B17 to B20 are approximately 3–10%. (Reproduced from Ref. [37])

β-rhombohedral boron has been found to transition from a nonmetal to a superconductor under a high pressure of 160 GPa at a temperature of 4 K [36]. The transition temperature is 6 K below 175 GPa and 11 K below 250 GPa.

6.4.3.2 Self-Compensation Properties

As described above, the β-rhombohedral boron is a semiconductor with about 1.5 eV band gap. There is an intriguing electronic property, the self-compensation, that only occurs in this material among the crystalline elemental semiconductors [37]. Recalling the crystalline structure, as shown in Fig. 6.15(a), the atomic arrangement is composed of (b) four B_{12} icosahedral clusters at the corners and edge centers of the rhombohedral unit cell, (c) a single B atom at the center of the unit cell, and two B_{28} clusters arranged symmetrically relative to the center atom along the main body diagonal. Ideally, chemical formula of the β-rhombohedral boron is $(B_{12})_4(B_{28})_2B$, i.e., B_{105}; however, the actual formula has known to be $B_{106.6}$ [9]. The discrepancy is understood by the self-compensation of carriers in the β-rhombohedral boron. The electron deficiency of the B_{12} cluster and the electron excess of the B_{28} cluster are self-compensated by structural defects (interstitial B atoms at the B16 to B20 sites and a vacancy at the B13 site, respectively). The defect sites are indicated in Fig. 6.15. The self-compensation occurs even after electron doping by external impurities, such as Li or Mg atoms, by removing interstitial B atoms at the B16 site and the generation of vacancies at the B13 and B4 sites. The β-rhombohedral boron keeps the semiconducting nature even by high concentration of Li as shown in Figs. 6.12(a) and 6.16.

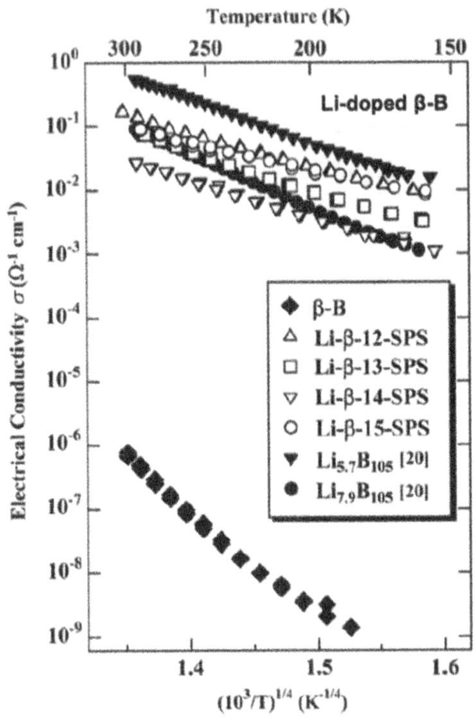

Fig. 6.16 Temperature dependence of electrical conductivity of Li-doped β-rhombohedral boron [37]. Li-β-x-SPS means Li_xB_{105} sintered sample using spark plasma sintering (SPS). The reference number [20] corresponds to [30] in this article and shown in Fig. 6.12(a). (Reproduced from Ref. [37])

6.4.3.3 Photoconduction

When steady light irradiation corresponding to the excitation from the valence band to the conduction band is performed, photoconduction increases over a long time, and it takes more time to reach a steady state the lower the temperature as shown in Fig. 6.17. Although Fig. 6.17 shows the case of irradiation with 1.5 eV light, there is a similar phenomenon at other wavelengths, which makes it difficult to measure photoconductivity in boron [38]. Furthermore, an oscillating current with a period exceeding 1 min has been observed, which becomes unstable when a high electric field of 2 kV/cm or higher is applied at a low temperature of 220 K or less, which is similar to the switching phenomenon in electrical conduction [39].

A further hysteresis phenomenon was observed in photoconduction, and the results are shown in Fig. 6.18. When irradiating 1.3 eV light at 286 K, it takes about 3 hours for the photocurrent value to reach a steady value. Figure 6.18 shows a photocurrent decay curve measured immediately after the light irradiation was stopped. It can be seen that it is composed of a component that decays rapidly in about 10 seconds (A process) and a component that decays slowly after 3 hours (B process). Figure 6.19 shows the result of measuring the thermal stimulation current when irradiating 1.3 eV light at 167 K for 3 hours and raising the temperature at a constant speed after stopping the irradiation. Two peaks are observed at about

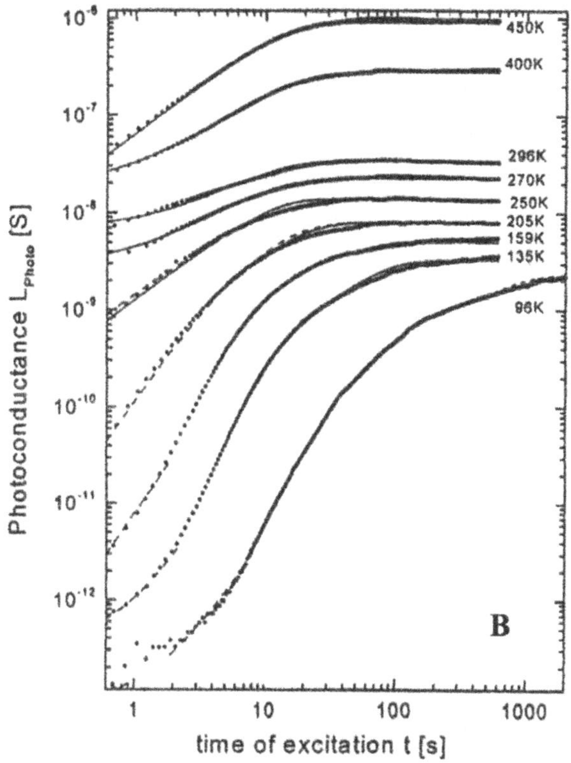

Fig. 6.17 Time dependence of the photoconductivity of β-rhombohedral boron exposed to 1.5 eV steady light irradiation [38]. (By courtesy of H. Werheit)

200 K and 300 K. This indicates that there are at least two types of electron traps. Therefore, Fig. 6.18 shows a rapid decay component that is thermally excited in the conduction band and immediately recombines and disappears from a shallow trap corresponding to the 200 K pea, and a slow decay component that is slowly thermally excited from a deep trap [40].

Figure 6.20 shows a spectrum of photoconduction caused by weak light irradiation at room temperature. The 0.8 μm (1.55 eV) peak is thought to be due to excitation from the valence band to the conduction band. The 5.64 μm (0.22 eV) peak may be due to excitation from the valence band to the acceptor level. The transition involved in the 0.35 μm (3.54 eV) peak is unknown [29].

It has been reported that, if electrons are accumulated in traps by pre-excitation prior to photoconductivity measurement with the fundamental absorption light at 155 K, the spectral structure of electrons trapped in traps appears at the bottom of the low-energy side of the fundamental absorption measured at 90 K. However, at 320 K, pre-excitation reduces the photoconduction instead. The results are shown in Fig. 6.21 [41].

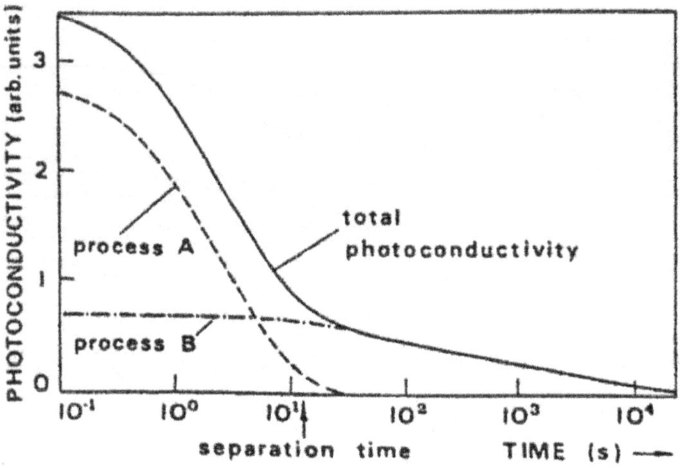

Fig. 6.18 Attenuation curve after stopping irradiation of the photoconductivity of β-rhombohedral boron saturated by irradiation with 1.3 eV of steady light for 3 hours at room temperature [40]. The figure shows that the fast decay A process coexists with the slow decay B process. (Reprinted by permission from 11th Int. Symp. Boron, Borides and Related Compounds 1993)

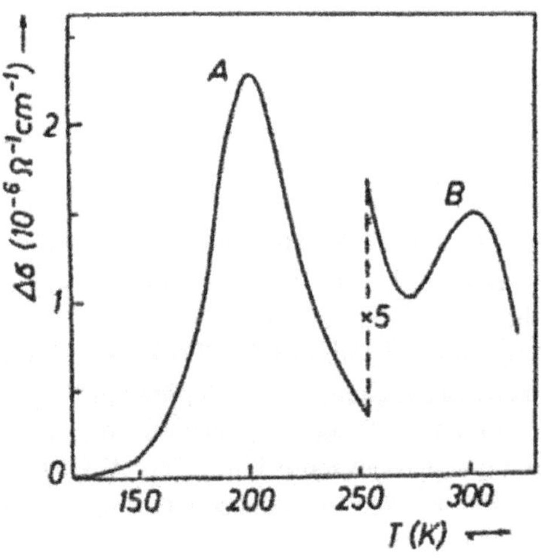

Fig. 6.19 Temperature dependence of thermally stimulated current when 1.3 eV light is irradiated for 3 hours at 167 K, and the temperature is raised at a constant speed after irradiation is stopped [40]. (Reprinted by permission from 11th Int. Symp. Boron, Borides and Related Compounds 1993)

6.4.3.4 Dielectric Constant

The reported DC dielectric constants are 10.0 ± 0.5 [42], 10.6 [43], and 11.55 [44] and drop to 8 ± 0.5 at 9.5 GHz [45].

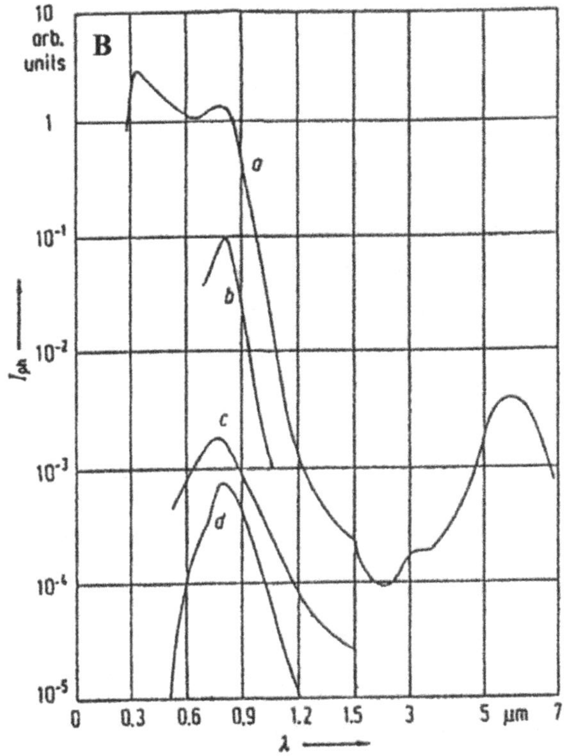

Fig. 6.20 Photoconductivity spectrum of β-rhombohedral boron when exposed to weak steady light at room temperature [29]. The vertical axis is an arbitrary unit. For ease of viewing, the vertical axis of each graph for different samples and measuring researchers is shifted. (**a**): Measurement with modulated light. (**b**, **c**, **d**): Measurement with steady light. (By courtesy of H. Werheit)

6.4.4 Optical Properties

6.4.4.1 Optical Absorption

Fig. 6.22 shows the absorption coefficient and reflectance data compiled by H. Werheit, and Fig. 6.23 shows the real and imaginary parts of the absorption coefficient, refractive index, and dielectric function of all regions calculated from these data [46]. Since the reflectance measurement ends at 40 eV, extrapolation to 700 eV where absorption measurements are made requires assumptions about dispersion. However, since dispersion based on classical theory is not practical, it is assumed that there is no dispersion. Although this assumption has a small effect on the imaginary part of the dielectric function, the real part which is constant in this energy region gives different results, depending on the assumption regarding dispersion.

In the light absorption spectrum of boron, similarly to other semiconductors, a region due to transition between energy bands and a region due to lattice vibration are clearly separated. The maximum absorption is the fundamental absorption due to inter-band transition. From 2.5 to 5.5 eV, an absorption coefficient larger than the graph shown here has also been reported. Absorption in the phonon region,

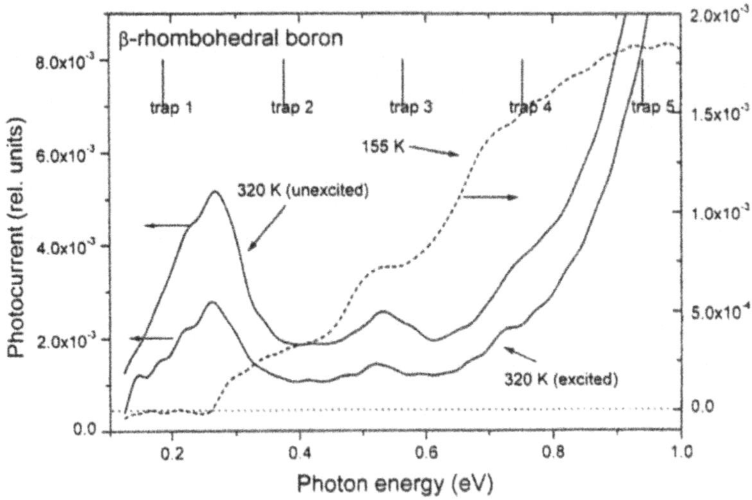

Fig. 6.21 Effect of preliminary excitation by fundamental absorption light on the photoconductivity of β-rhombohedral boron [41]. The 155 K data is the photoconduction caused by pre-excitation. Trap 1 to trap 5 indicate the positions of electron capture levels estimated by H. Werheit. (Reprinted by permission from R. Uno, K. Kimura, Chap. 1, Fundamentals and Applications of Boron, Borides and Related Materials (CMC Publishing Co., LTD. 2008). Copyright (2008) CMC Publishing Co.)

which occurs mainly with phonons in the regular icosahedron, is smaller than that of ion crystals. This indicates that the bond of β-rhombohedral boron is weak in ionic bonding and is mostly covalent bonds. In the region between above the two regions, the absorption is weak, and the absorption that depends on the individuality of the sample and impurities, particularly the carbon content, is observed. The high-purity sample shown here has a carbon content of only 66 to 90 ppm. In the low-energy region ($\leq$0.04 eV), the absorption depends strongly on the individuality of the sample. In high-purity samples, absorption decreases even beyond the phonon region, but in low-purity samples, absorption increases due to high-frequency conduction. In the far-infrared region, the dielectric function can be extrapolated to the DC dielectric constant, indicating that there is no observable dispersion [46].

Near the absorption edge, anisotropy due to the direction of the polarization plane is observed, and the results at room temperature are shown in Fig. 6.24. When the polarization plane is perpendicular to the c-axis, a fine structure is seen in the infrared region [47]. Similar to photoconduction, optical absorption has a hysteresis when measured at room temperature or lower, and the measured value is affected by light irradiation performed before the start of measurement. Therefore, a reproducible result may not be obtained. It has been found that the effects of this light irradiation can be eliminated by heating the sample to 100 °C or higher under dark conditions [48].

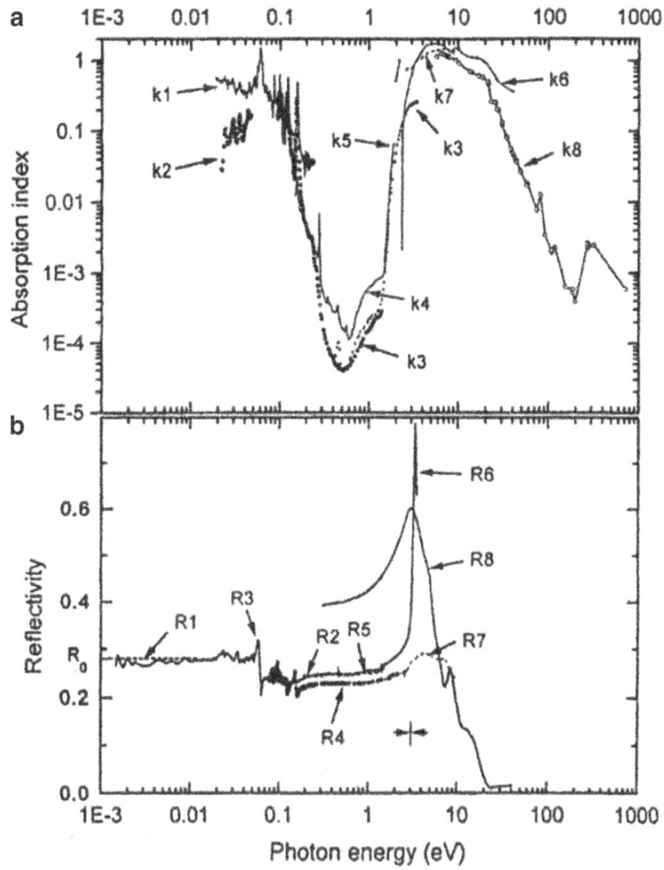

Fig. 6.22 Measurement data of optical absorption coefficient and reflectance of β-rhombohedral boron at room temperature [46]. (**a**) The subscripts of the absorption coefficient k and (**b**) the reflectance R indicate data from different references. (By courtesy of H. Werheit)

6.4.4.2 Photoluminescence

Photoluminescence is observed at low temperatures below about 100 K. Figure 6.25 shows the spectrum at 4.2 K for several excitation energies. The spectrum is broad but the peak position is 1.14 eV, which is considered to be the light emission due to recombination of the electrons in the first electron capture level and the holes in the intrinsic acceptor level described in "band structure" [49].

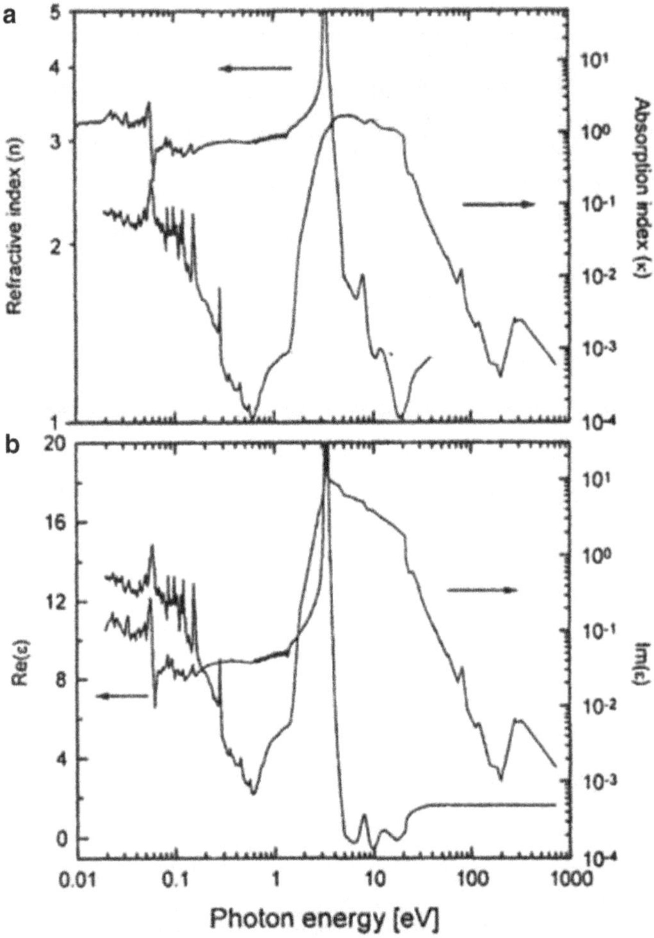

Fig. 6.23 Absorption coefficient and refractive index of β-rhombohedral boron at room temperature and real part and imaginary part of dielectric function [46]. (By courtesy of H. Werheit)

6.4.4.3 Infrared Absorption/Raman Scattering

Information on lattice vibration can be obtained from infrared absorption and Raman scattering, but according to group theory analysis, infrared active modes include A_{2u} 31 modes and E_u 52 modes, and the measurement results have a complex structure as shown in Fig. 6.26 [44]. Raman activity modes also include A_{1g} 31 modes and E_g 52 modes, and the Raman shift data shows the same structure as infrared absorption [50].

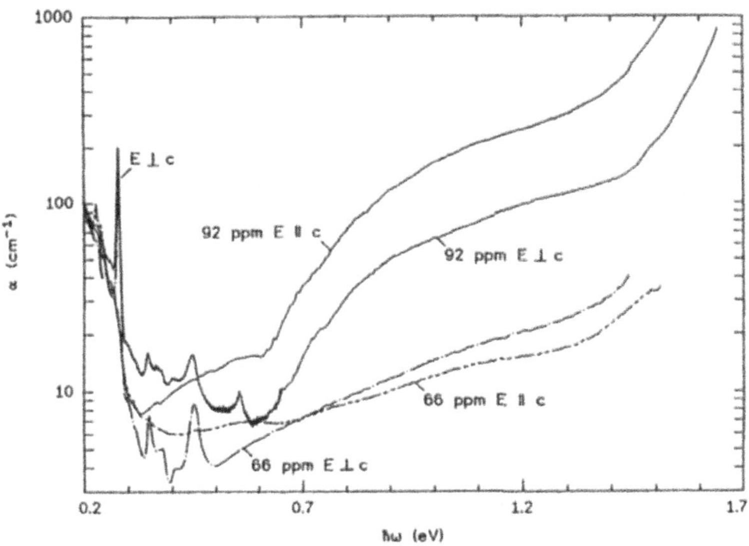

Fig. 6.24 Anisotropy of light absorption spectrum near the absorption edge of β-rhombohedral boron at room temperature [47]. Samples with 66 and 92 ppm carbon content are shown. (By courtesy of H. Werheit)

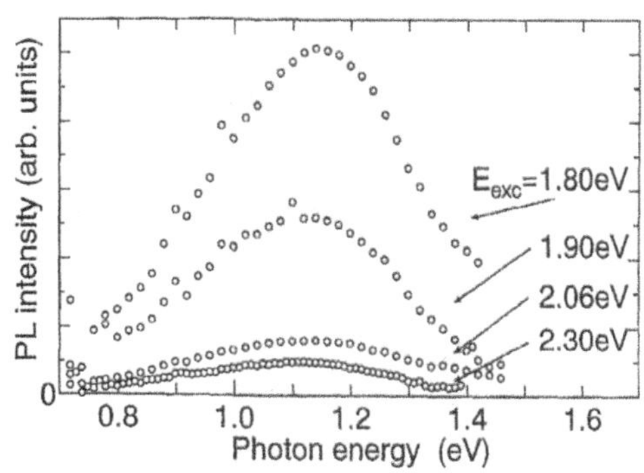

Fig. 6.25 Photoluminescence spectrum of β-rhombohedral boron at 4.2 K for several excitation energies [49]. (Reproduced from Ref. [49])

6.4.5 *Magnetic Properties*

6.4.5.1 Magnetic Susceptibility

Pure β-rhombohedral boron is a diamagnetic material and has a magnetic susceptibility of about -0.8×10^{-6} emu/gOe. Doping metal adds a temperature-independent Pauli paramagnetic χ_P and Landau diamagnetism χ_L, and temperature-dependent Curie-Weiss paramagnetism and ferromagnetism. If there is no ferromagnetism due to the doped metal or impurities, the magnetic susceptibility χ becomes

$$\chi(T) = \chi_0 + C/(T - \theta_P)$$

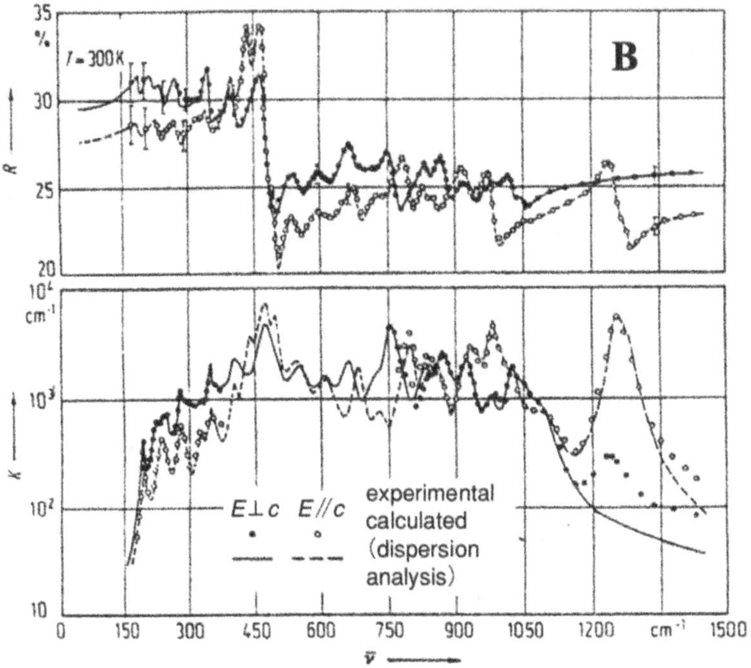

Fig. 6.26 Infrared absorption spectrum of β-rhombohedral boron at room temperature [44]. The reflectance R and the absorption coefficient K are shown. (By courtesy of H. Werheit)

where χ_0 is the temperature-independent component. $\chi(T)$ is measured by changing the doping amount, and the change of χ_0 depending on the doping amount and the Curie constant C can be obtained by parameter fitting of χ_0, θ_P, and C. The change in χ_0, $\Delta\chi_0$ is

$$\begin{aligned}\Delta\chi_0 &= \Delta\chi_P + \Delta\chi_L \\ &= {\mu_B}^2 \Delta N(E_F)\left\{1-(1/3)\left(m/m^*\right)^2\right\}\end{aligned}$$

Here, μ_B is a Bohr magneton, $\Delta N(E_F)$ is a change in density of state at the Fermi level E_F, and m and m* are the mass and effective mass of electrons. If m* is known, $\Delta N(E_F)$ can be obtained. Also, the effective Bohr magneton number can be calculated from C. These are used as one of the means of band structure research [30].

6.4.5.2 Magnetic Resonance

Electron spin resonance is used as one of the methods for obtaining information on the number of unpaired electrons and their states. The g factor obtained at room temperature is 2.003, indicating that the orbital angular momentum is small. The spin resonance signal gradually increases with the light irradiation of the GaAs diode and takes about 3 hours to reach a steady state. Attenuation after cessation of light irradiation is similar to photoconduction and consists of a rapid process that decays in about 10 seconds and a slow process that lasts about 3 hours [51].

6.4.6 *Mössbauer Effect*

The valence of a transition metal element doped in the β-rhombohedral boron is important for understanding the function of the element. Measurements were made using the fact that the valence of Fe ions is determined by the Mössbauer effect. Fe atoms are doped at A1 site and D site, but the Mössbauer effect was measured for Fe-doped and Al and Fe 2 elements-doped samples using the fact that Al atoms are more easily doped into the A1 site than Fe atoms. As a result, it was found that Fe at the A1 site was Fe^{3+} and Fe at the D site was Fe^{2+} [28].

6.5 Band Structure of β-Rhombohedral Boron

The band structure proposed by H. Werheit is very complex [38], and the existence of six electron capture levels distributed at almost equal intervals in the band gap is inconsistent with the existence of at least two groups of electron capture levels indicated by thermally stimulated current data. The simplified model adopted to explain photoluminescence only assumes two types of electron capture levels. Furthermore, since fast transition recombination seems difficult to achieve in cascade-type recombination through six electron capture levels, which is considered as the recombination process, it is thought that future research is required. Therefore, the previously proposed band structure [29] is slightly modified to have a structure in which two electron capture level groups and one acceptor level group are placed in the band gap. The band gap value is 1.56 eV, with the intrinsic acceptor level at 0.2 eV above the valence band and the first electron capture level at 0.2 eV below the conduction band. Although it is difficult to determine the energy position of the electron capture level only from the peak of the thermally stimulated current, the second electron capture level is estimated to be around 0.35 eV below the conduction band [52]. The electron capture level is created below the conduction band due to electron-lattice interaction, and the acceptor level is separated from the valence band by the Jahn-Teller effect of the B_{12} icosahedron and defects. A specific band structure diagram is shown in Fig. 6.27. Even in the latest first-principles calculation [53], the band gap is 1.4 eV and the position of the intrinsic acceptor level is 0.35 eV above the valence band, which is not completely consistent with experimental results.

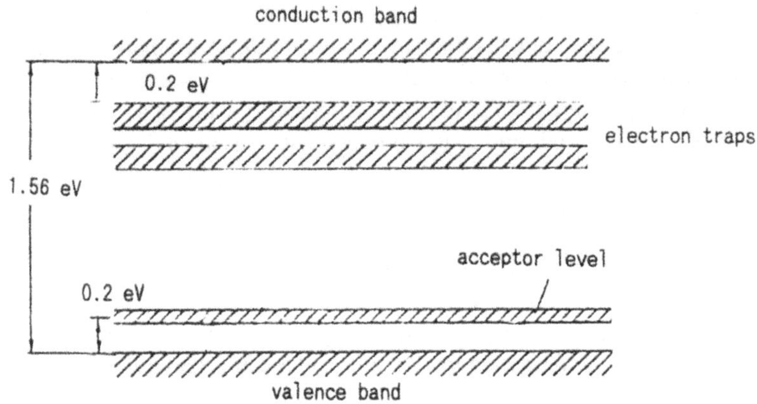

Fig. 6.27 Band structure of β-rhombohedral boron. (Reprinted by permission from R. Uno, K. Kimura, Chap. 1, Fundamentals and Applications of Boron, Borides and Related Materials (CMC Publishing Co., LTD. 2008). Copyright (2008) CMC Publishing Co.)

In terms of its electrical properties, β-rhombohedral boron is a unique substance that exhibits a hysteresis phenomenon with an unusually long relaxation time. It is doubtful whether this peculiarity can be explained by the high density of electron capture levels and acceptor levels and the strong localization of their wave functions. In addition, future research is required to determine the characteristics that give rise to the features of B atoms, such as their complex crystalline structure, atomic positions with a filling rate that is not 100%, and wide variety of doping sites for metal atoms.

6.6 Physical Properties of Other Crystals

6.6.1 α-Rhombohedral Boron

α-Rhombohedral boron B_{12} has the simplest structure among boron crystals and has attracted attention as an academic research subject. In particular, it is theoretically predicted that a large density of states in the Fermi level, which is particularly related to superconductivity, may be obtained by doping with Li [54], and experiments aiming at superconductivity are being carried out [55]. The critical temperature of superconductivity about 7 K is realized for $Li_{0.3}B_{12}$ [56].

α-Rhombohedral boron is made by heating amorphous boron at around 1200 °C for a long period, but no large crystals are formed. It is red at room temperature and has a relatively large electrical conductivity ($10^{-3}(\Omega{\cdot}\mathrm{cm})^{-1}$) as a semiconductor, and mobility is 120 $\mathrm{cm^2/V{\cdot}s}$. However, it easily turns into β-rhombohedral boron at high temperatures and is unstable, and applications have not been developed [57].

Boron carbide $B_{12}C_3$, which has a three-atom chain of C-B-C in the center void of the rhombohedral unit cell of α-B_{12}, completely compensates for deficient electrons in the valence band and is an example of one stable boride system. $B_{12}C_3$ is relatively stable even at high temperatures, has a large Seebeck coefficient, and is promising as one type of thermoelectric generating element. $B_{12}P_2$ and $B_{12}O_2$ containing a P-P atom chain or two O atoms instead of a C-B-C chain are also stable boron compounds.

6.6.2 Amorphous

Since the radial distribution function obtained from X-ray diffraction measurement is very similar to that calculated from the structure of β-rhombohedral boron, the local structure is considered to be close to β-rhombohedral boron [58]. The electrical conductivity at room temperature is $10^{-4}(\Omega\text{cm})^{-1}$, and the Seebeck coefficient is 420 μV/K [59]. The drift mobility is obtained by transient photoconductivity at 250 K, and the electron mobility is 2 x 10^{-1} cm^2/V·s and the hole mobility is 2 x 10^{-2} cm^2/V·s. Both show activation energy, which is thought to be due to hopping conduction [35].

6.6.3 Quasicrystal and Approximant

A quasicrystal is a solid that has an ordered structure but lacks periodicity. Since it is neither crystal nor amorphous, it has attracted interests not only in condensed matter physics but also in mathematics. Then, there have been discoveries of various quasicrystals but only in metallic alloy systems. The extensive research in boron-rich materials was motivated to synthesize semiconducting quasicrystals in covalent bonding system. This is based on the fact that varieties of crystalline structures with icosahedral clusters (B_{12}) are semiconducting and some of them have a unit cell that is almost as same as the prolate rhombohedron, which is one of the two unit cells in the three-dimensional Penrose pattern (the model of icosahedral quasicrystal) [60, 61]. Fig. 6.28 shows example of the two unit cells of icosahedral quasicrystal, (a) prolate and (b) oblate rhombohedra, which are theoretically constructed by analyzing β-rhombohedral boron [62, 63]. The icosahedral B_{12} cluster, the main structural unit in boron crystals, is stable enough to form various types of structure based on icosahedral networks. The α- and β-rhombohedral boron, in fact, show direct relevance to the quasicrystal structure, i.e., they are the 0/1–2/0 rhombohedral crystalline approximants of the icosahedral quasicrystal [64, 65]. Furthermore, theoretical investigations have also been carried out in boron compounds, such as boron-carbon materials and metal boride structures [64], and the 1/0–1/0–0/1

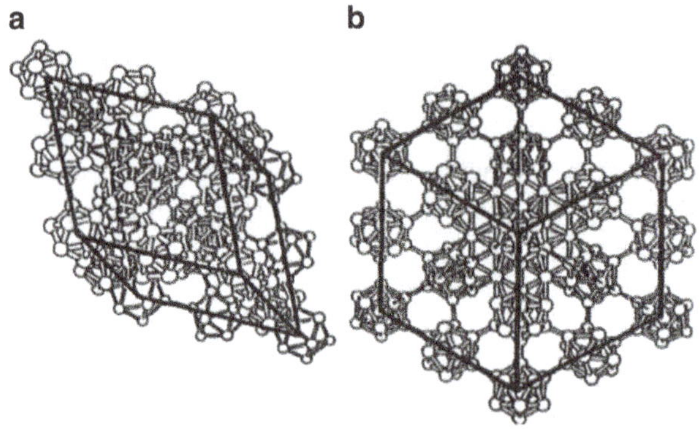

Fig. 6.28 Structure models of two unit cells of the icosahedral quasicrystal. (**a**) prolate and (**b**) oblate (**b**) rhombohedra, constructed from the unit cell of β-rhombohedral boron [62]. Reproduced from Ref. [62]

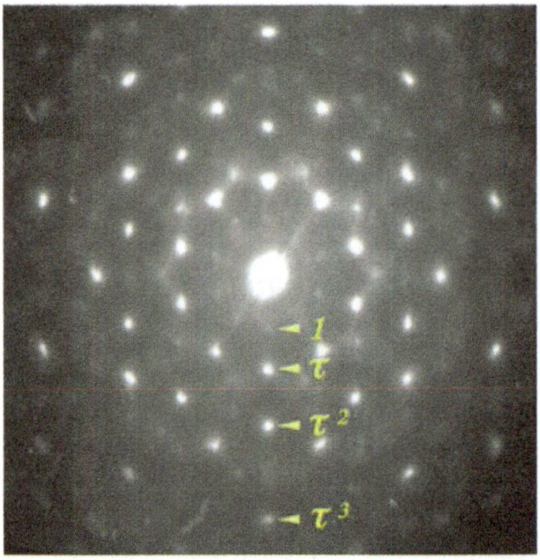

Fig. 6.29 Electron diffraction pattern of the rapidly quenched $B_{40}Ti_{12}Ru_{48}$ alloy [66]. A distance from the center spot can be scaled by the golden mean $\tau = (1 + \sqrt{5})/2$. (Reproduced from Ref. [66])

orthorhombic crystalline approximant, which is higher approximant than the 0/1–2/0 rhombohedral one, was found in B-several at.% C system as a metastable phase [63].

The first quasicrystalline boride was found in the $B_{40}Ti_{12}Ru_{48}$ alloy, formed by the rapid solidification [66)]. As shown in Fig. 6.29, the diffraction pattern clearly reveals a tenfold rotational decagonal symmetry. The decagonal structure is described in terms of periodically stacked quasiperiodic atomic planes. A quasicrystal now becomes one of the boron compound forms, but the other examples are not yet discovered.

6.7 Chemical Properties of Boron Solids

Boron, like silicon, is inert in air at room temperature. It is insoluble in water, hydrochloric acid, and hydrofluoric acid and is also hardly soluble in concentrated nitric acid, concentrated sulfuric acid, and molten alkali. However, it reacts with many elements at high temperatures to produce many unique compounds.

When molten boron is cooled and solidified, it becomes β-rhombohedral boron, which is stable at higher temperatures. α-Rhombohedral boron is more stable at 1200 °C or lower and can be formed by heating amorphous boron to 1200 °C for a long period.

Stable B_2O_3 is very difficult to crystallize, which is thought to be due to the boron atom. Boron is sometimes used as an additive when making glassy materials.

6.7.1 Boron Valence

From the electron configuration of boron atoms, the valence is thought to be trivalent, producing hydrides, fluorides, oxides, phosphides, arsenides, etc.

There is a unique group of compounds with nonmetallic elements that should be called B_4C compounds. In B_4C, α-rhombohedral boron has a large void in the center of the unit cell, C-B-C chains are arranged in the rhombohedron diagonal direction (c-axis of hexagonal coordinates), and an extra C replaces one B atom in the adjacent icosahedron. In this system, when B increases according to the number ratio of B and C, a part of C-B-C changes to C-B-B, and $B_{13}C_2$ becomes a complete C-B-B chain. When B further increases, the C replacing the icosahedron B decreases, and the compound dissolves to $B_{9.3}C$. In the void of α-rhombohedral boron, two atoms of phosphorus, arsenic, oxygen, or sulfur can enter adjacently on the diagonal line, resulting in B_6P, B_6As, B_6O, or B_6S. In this case, the distance between O-O and S-S is relatively large, and it is thought that there is no bond between the two atoms.

Boron has a high electronegativity among group 13 (IIIb) elements and forms a wide variety of compounds with metal elements. In metal borides, if the number ratio of metal atoms to boron atoms is R, R takes a wide range of values from 16/1 of $Nd_2Fe_{14}B$, a permanent magnet material, to 1/66 of YB_{66}. B tends to form short B-B bonds, and when $R \geq 3$ with many metal atoms, B becomes arranged in an isolated state in the lattice formed by the metal atoms. With R of around 3/2, B-B diatomic groups are formed. When $R = 1$, B forms a one-dimensional chain, and as B increases, a side chain is formed. When R is around 5/6, a double-chain structure is formed. When R becomes even smaller, B forms a two-dimensional layered structure like graphite, and when R is around 1/2, B forms a crystal in which the B layer and the metal atomic layer are stacked. When R is around 1/4, B begins to form a three-dimensional structure, and when R is around 1/6, B forms a three-dimensional structure in which cubes are stacked with an octahedron at the vertex, with metal atoms in the void. When R is around 1/12, metal atoms are incorporated into the icosahedron formed by 12 Bs or the three-dimensional structure based on a cubic octahedron. At R < 1/12, it can be considered that the metal atoms have been incorporated into the void of the icosahedron and the three-dimensional boron structure consisting of a B_{20} unit, B_{22} unit, B_{28} unit, etc. that unite half of the pentagonal pyramid.

In addition to the above, many borides, such as hydrogen compounds and organoboron compounds having a three-center bond, have been recognized. Considering boron in these boron compounds as an ion, its valence is very complex, and it is necessary to extend the concept of structural chemistry to the properties of the three-dimensional structures created by boron in order to understand it [67].

6.7.2 Reaction with Hydrogen

Heated α-rhombohedral or β-rhombohedral boron and molten boron adsorb a large amount of hydrogen. Solid boron reacts slowly with hydrogen at 1000 °C or higher to form many hydrides such as diborane (B_2H_6) [68].

6.7.3 Reaction with Halogen Elements and Halides

All halogen elements react with solid boron to form boron halide. Fluorine reacts at room temperature, chlorine and bromine react at 400 °C or higher, and iodine reacts at 1000 °C or higher. Boron halide reacts with solid boron at 1000 °C or higher. For example, BCl_3 becomes subchloride BCl by the following reaction [69].

$$2B\,(solid) + BCl_3\,(gas) \rightarrow 3BCl\,(gas)$$

6.7.4 Reaction with Oxygen and Oxides

β-Rhombohedral boron forms a thin film of B_7O or B_6O exhibiting p-type conductivity on the surface in oxygen at 500 °C and 10^{-4} mbar. Therefore, electrical conduction increases, thermoelectric power decreases, and light absorption increases [70]. When crystalline boron is heated in air, it becomes coated with a B_2O_3 film at 600–1000 °C, and at 1000 °C or higher, B_2O_3 evaporates and oxidation proceeds. Amorphous boron is oxidized more intensely than crystalline boron, and BN is also formed during the oxidation process at 1100–1300 °C in air [71].

At high temperatures, boron reacts with B_2O_3 to form suboxide $(BO)_2$, and B_6O is formed at 1500 °C. The reaction formula is as below [72].

$$16B + B_2O_3 \rightarrow 3B_6O$$

Boron reacts with molten SiO_2 to form Si, SiB_2, and SiB_4. NO and N_2O react with boron to form B_2O_3 and BN. At 1000 °C or above, boron reacts with transition metal oxides to form B_2O_2 and transition metal borides [73]. The reaction is as below.

$$MO_2 + 4B \rightarrow MB_2 + B_2O_2 \quad (M = Ti, Zr, Hf)$$

6.7.5 Reaction with Sulfur, Sulfides and Selenium

Boron reacts with sulfur at high temperatures to form glassy B_2S_3. Furthermore, boron reacts with B_2S_3 to produce $(BS)_2$ [74].

Boron reacts with H_2S at 700 °C and 1 atm to form HBS_2. HBS_2 decomposes into glassy B_2S_3 and H_2S at 300 °C and below [75]. Low-pressure H_2S reacts with boron at 1150–1300 °C to form gaseous HBS and H_2 [76]. Metal sulfides, such as Cr_2S_3, V_2S_3, and Ti_2S_3, react with amorphous boron and β-rhombohedral boron at 700–1000 °C to form the metal borides CrB_2, VB_2, and TiB_2 [77].

Boron reacts with selenium at 850–950 °C to form B_2Se_3 [78].

6.7.6 Reaction with Nitrogen, Phosphorus, and Arsenic

Nitrogen reacts with α-rhombohedral and β-rhombohedral boron at 1200–1600 °C. In the reaction with α-rhombohedral boron, BN is formed. In the reaction with β-rhombohedral boron, B_6N is formed first, and B_6N reacts with nitrogen to form BN.

Boron reacts with gaseous phosphorus at 1000 °C to form BP, $B_{14}P$, and $B_{12}P_2$. It is thought that BP is produced first and then decomposes to become stable $B_{12}P_2$ [79].

Boron reacts with arsenic at 800 °C to form BAs, and at higher temperatures, BAs decompose into $B_{12}As_2$ [80].

6.7.7 Reaction with Metals

The bond between boron and metals is generally weaker than the bond with nonmetals. Borides are mostly made by reacting directly with metals. More than 200 borides have been made, and their molecular formulas extend from MB_{100}, a metal-doped boron, to M_6B. Common borides are MB, M_3B_4, MB_2, MB_4, MB_6, and MB_{12} [74].

In order to dope metals into boron crystals with many voids, one method is vapor diffusion, where boron crystals heated to around 1000 °C are exposed to metal vapor for a long time and metal atoms are diffused from the crystal surface. The second method is where metal is added to molten boron and solidified and then made uniform using a zone melting method. Although many metals, such as alkali metals, alkaline earth metals, and transition metals, can be doped, there are differences in the amount that can be doped by the metal, and some metals have limits in terms of the voids that can be doped. For example, in β-rhombohedral boron, Mg partially occupies D, E, F, and H sites simultaneously, Zr partially occupies D and E sites, and V almost completely occupies only A1 sites (see "Physical Properties").

The reactivity between boron and alkali metals decreases as the atomic weight of the metal increases. The temperatures for complete reaction are 700 °C for Li, 900 °C for Na, and 1100 °C for K. When the particle size is the same, α-rhombohedral boron is more reactive than β-rhombohedral boron [81]. Binary borides or ternary borides are formed by adding rare earth elements or transition elements, etc. to boron melted by arc or argon plasma [82].

References

1. E. Amberger, K. Ploog, J. Less-Common Metals **23**, 21 (1971)
2. K. Ploog, ibid **31**, 15 (1973)
3. A.R. Oganov, J. Chen, C. Gatti, Y. Ma, Y. Ma, C.W. Glass, Z. Liu, T. Yu, O.O. Kurakevych, V.L. Solozhenko, Nature **457**, 863 (2009)
4. K. Shirai, Jpn. J. Appl. Phys. **56**, 05FA06 (2017)
5. W. Hayami, S. Otani, J. Phys. Chem. C **111**, 688 (2007)
6. Z. Wang, Y. Shimizu, T. Sasaki, K. Kawaguchi, K. Kimura, N. Koshizaki, Chem. Phys. Lett. **368**, 663 (2003)
7. G.T. McConville, D.B. Sullenger, R.E. Zielinski, D.U. Gubser, Phys. Letter **58 A**, 257 (1976)
8. M. Fujimori, T. Nakata, T. Nakayama, E. Nishimori, K. Kimura, M. Takata, M. Sakata, Phys. Rev. Lett. **82**, 4452 (1999)
9. G.A. Slack, C.I. Hejna, M.F. Garbauskas, J.S. Kasper, J. Solid State Chem. **76**, 52 (1988)
10. J.L. Hoard, R.E. Hughes, D.E. Sans, J. Am. Chem. Soc. **80**, 4507 (1958)
11. M. Vlasse, R. Naslain, J.S. Kasper, K. Ploog, J. Solid State Chem. **28**, 289 (1979)
12. E. Nishibori, H. Hyodo, K. Kimura, M. Takata, Solid State Sci. **47**, 27 (2015)
13. M. Takeda, K. Kimura, Parity **12**(3), 8 (1997) in Japanese
14. K. Kimura, M. Takeda, M. Fujimori, R. Tamura, H. Matsuda, R. Schmechel, H. Werheit, J. Solid State Chem. **133**, 302 (1997)
15. Y. Takagiwa, K. Kimura, Sci. Tech. Adv. Mater. **15**, 044802 (2014)
16. H.C. Lonquet-higgins, M.D.V. Robert, Proc. Roy. Soc. London **A230**, 110 (1955)
17. M. Fujimori, K. Kimura, J. Solid State Chem. **133**, 178 (1997)
18. R. Franz, H. Werheit, AIP Conf. Proc. **231**, 29 (1991)
19. T. Lundstrom, AIP Conf. Proc. **140**, 19 (1986)
20. V. Adasch, K.-U. Hess, T. Ludwig, N. Vojteer, H. Hillebrecht, J. Solid State Chem. **179**, 2900 (2006)
21. S. Anderson, T. Lundstoem, J. Solid State Chem. **2**, 603 (1970)

22. I. Higashi, M. Kobayashi, Y. Akagawa, K. Kobayashi, J. Bernhard, AIP Conf. Proc. **231**, 224 (1991)
23. M. Kobayashi, I. Higashi, H. Matsuda, K. Kimura, J. Alloys Comp. **221**, 120 (1995)
24. U. Kuhlmann, H. Werheit, et al., J. Alloys Comp. **186**, 187 (1992)
25. H. Hyodo, S. Araake, S. Hosoi, K. Soga, Y. Sato, M. Terauchi, K. Kimura, Phys. Rev. B **77**, 024515 (2008)
26. H. Werheit, K. De Groot, W. Malkemper, T. Lundstroem, J. Less-Common Metals **82**, 163 (1981)
27. H. Matsuda, N. Tanaka, T. Nakayama, K. Kimura, Y. Murakami, H. Suematsu, M. Kobayashi, I. Higashi, J. Phys. Chem. Solids **57**, 1167 (1996)
28. T. Nakayama, H. Matsuda, K. Kimura, H. Ino, J. Solid State Chem. **133**, 342 (1997)
29. H. Werheit, *Landolt-Boernstein, Numerical Data and Functional Relationship in Science and Technology*, vol 17 e (Springer, 1983), p. 9
30. H. Matsuda, T. Nakayama, K. Kimura, Y. Murakami, H. Suematsu, M. Kobayashi, I. Higashi, Phys. Rev. **B52**, 6102 (1995)
31. R. Schmechel, H. Werheit, J. Solid State Chem. **133**, 335 (1997)
32. H. Ozawa, R. Takei, R. Uno, J. Less-Common Metals **67**, 263 (1979)
33. H. Kim, K. Kimura, Mater. Trans. **52**, 41 (2011)
34. O.A. Golikova, A.Y. Kiskachi, T. Khomidor, Sov. Phys. Semicond. (English Translation) **4**, 683 (1970)
35. M. Takeda, K. Kimura, K. Murayama, J. Solid State Chem. **133**, 201 (1997)
36. M.I. Eremets, V.V. Struzhkin, R. Ho-Kwang Mao, J. Hemley, Science **293**, 272 (2001)
37. H. Hyodo, A. Nezu, K. Soga, K. Kimura, Solid State Sci. **14**, 1578 (2012)
38. R. Schmechel, H. Werheit, J. Solid State Chem. **154**, 68 (2000)
39. O.A. Golikova, M.M. Kazanin, E.L. Lutsenko, J. Less-Common Metals **82**, 177 (1982)
40. A.J. Nadolny, Jpn. J. Appl. Phys., Seires **10**, 46 (1994)
41. H. Werheit, "Electric refractory materials, p 589", ed. Y. Kumashiro, Marcel Dekker; H. Werheit, M. Schmidt, R. Schmechel, T. Lundstroem, J. Solid State Chem., 154, 93 (2000)
42. W. Neft, K. Seiler, Boron prep. Properties Appl. Papers Intern. Symp. Paris 1964 **2**, 143 (1965)
43. J. Lagrenaudi, J. Chem. Phys. **50**, 629 (1953)
44. H. Binnenbruck, H. Werheit, Z. Naturforsch., 34 a, 787 (1979)
45. W.P. Lonc, J. Less-Common Metals **82**, 149 (1982)
46. H. Werheit, R. Schmechel, J. Solid Stat. Chem. **133**, 129 (1997)
47. H. Werheit, A. Hausen, H. Binnenbruck, Phys. Status Solidi (b) **176**, 415 (1993)
48. H. Werheit, R. Franz, J. Less-Common Metals **177**, 163 (1986)
49. A. Hori, T. Tada, K. Kimura, J. Phys, Soc. Jpn. **67**, 4279–4282 (1998)
50. H. Binnenbruck, H. Werheit, J. Less-common Metals, 47, 91 (1976)
51. A.J. Nadolny, Phys. Stat. Sol. (b) **65**, 801 (1974)
52. M. Prudenziati, G. Majni, A. Alberigi-Quaranta, J. Phys. Chem. Solids **33**, 245 (1972)
53. M.J. van Setten, M.A. Uijttewaal, G.A. de Wijs, R.A. de Groot, J. Am. Chem. Soc. **129**, 2458 (2007)
54. S. Gunji, H. Kamimura, Phys. Rev. **B54**, 13665 (1996)
55. K. Soga, A. Oguri, S. Araake, M. Terauchi, A. Fujiwara, K. Kimura, J. Solid State Chem. **177**, 498 (2004)
56. T. Nagatochi, H. Hyodo, A. Sumiyoshi, K. Soga, Y. Sato, M. Terauchi, F. Esaka, K. Kimura, Phys. Rev. B **83**, 184507 (2011)
57. O.A. Golikova, N.E. Solovev, Y.A. Ugai, V.A. Feigelman, J. Less-Common Metals **82**, 362 (1982)
58. M. Kobayashi, I. Higashi, T. Oyama, N. Kanekama, M. Takami, K. Shirai, Jpn. J. Appl.Phys., Series **10**, 25 (1994)
59. A.A. Berezin, O.A. Golikova, M.M. Kazanin, T. Khomidov, D.N. Mirlin, A.V. Petrov, A.S. Umarov, V.K. Zaitsev, J. Non-Cryst. Solids **16**, 237 (1974)
60. I. Higashi, in *Boron-Rich Solids (AIP Conf. Proc. 140)*, ed. by D. Emin, T. Aselage, C. L. Beckel, I. A. Howard, C. Wood, (AIP, New York, 1986), pp. 1–10

61. K. Kimura, Y. Takamura, H. Yamashita, H. Ino, in *Boron-Rich Solids (AIP Conf. Proc. 231)*, ed. by D. Emin, T. Aselage, A. C. Switendick, B. Morosin, C. L. Beckel, (AIP, New York, 1991), pp. 528–531
62. M. Takeda, A. Hori, H. Yamashita, K. Kimura, Mater. Trans. JIM **34**, 128 (1993)
63. M. Takeda, K. Kimura, A. Hori, H. Yamashita, H. Ino, Phys. Rev. B **48**, 13159 (1993)
64. K. Kimura, Mater. Sci. Eng. **B19**, 67 (1993)
65. K. Kimura, A. Hori, H. Yamashita, H. Ino, Phase Transit. **44**, 173 (1993)
66. Y. Miyazaki, J. Okada, E. Abe, Y. Yokoyama, K. Kimura, J. Phys. Soc. Jpn. **79**, 073601 (2010)
67. D. B. Sullenger, C. H. L. Kennard, Scientific American (1996)
68. L. Barton, D. Nicholls, J. Inorg. Nucl. Chem. **28**, 1367 (1966)
69. V.A. Krenev, V.M. Bunin, V.I. Evdokimov, Izv. Akad. Nauk SSSR Neorgan. Materialy **6**, 1052 (1970)
70. H. Werheit, P. Runow, H.G. Leis, Phys. Stat. Sol **2**, K 125 (1970)
71. H.F. Rizzo, Boron Syn, Stract. Prop. Proc. Conf. Asbury Park N. J. 1959 **175** (1960)
72. P. Blackburn, A. Buchler, J. Staffer, J. Phys. Chem. **70**, 2469 (1969)
73. L. Barton, D. Nicholls, J. Inorg. Nucl. Chem., 28, 1367 (1966); R. Favre, F. Thevenot, Bull. Chim. France, p 3911 (1971)
74. J. Cueilleron, F. Thevenot, *Boron and Refractory Borides* (Springer, Berlin, 1977), p. 203
75. F.T. Greene, P.W. Gilles, J. Am. Chem. Soc. **86**, 3964 (1964)
76. R. Kirk, P. Timms, Chem. Commun. **18/9** (1967)
77. J. Cueilleron, G. Lahet, F. Thevenot, R. Paris, J. Less-Common Metals **24**, 317 (1971)
78. "Borverbindungen" Pt. 3, p 76, Springer, Berlin (1975)
79. "Borverbindungen" Pt. 3, p 93, Springer, Berlin (1975)
80. "Borverbindungen" Pt. 3, p 144, Springer, Berlin (1975)
81. R. Naslain, J. Etourneau, P. Hagenmuller, Electron. Technol. **3**, 35 (1970)
82. Y.B. Kuz'ma, N.S. Bilonizhko, S.I. Mykhalenko, G.F. Stepanchikova, N.F. Chavan, J. Less-Common Metals **67**, 51 (1979)
83. U. Häussermann, A.S. Mikhaylushkin, Inorg. Chem. **49**, 11270 (2010)

Index

A
Ag(110), 61–62
Ag(111), 9, 19, 34–38, 40, 41, 52–67, 74–79, 82, 84
Al(111), 20, 34, 62–64, 66, 112
α-rhombohedral boron, 2, 121–123, 128–131, 150–151, 153–156
α-tetragonal boron, 2, 121, 122, 125–126
Amorphous solid, 151, 153
Au(111), 34–36, 64–67
Aufbau principle, 5, 31, 32

B
β-rhombohedral boron, 2, 82, 121–125, 131–156
β-tetragonal boron, 2, 121, 122, 126–128
Boraphene, v, 9, 10, 30
Boron, v, vi, 1–22, 27–46, 51–56, 58–70, 73, 75–79, 81, 82, 84, 85, 89–99, 101, 102, 107–114, 121–156
Boronene, v
Boron hydride, 96
Boron nitride (BN), 73, 107–109, 155
Borophane, 89–106
Borophene, v, vi, 3–5, 9, 14, 15, 18–21, 27–46, 51–70, 73–86, 89–114
Borospherenes, 30, 31

C
Cluster, 7, 21, 32, 128–131
Cu(111), 19, 35, 38, 40, 43, 45, 52, 65–69

D
Dirac cones, 42, 62, 76–80, 82, 85, 92, 93
Dirac nodal lines, 77, 79, 82
Dirac nodal loops, 100

E
Electrical properties, 132–143, 150

F
Fullerene, 1, 5, 6, 29, 31, 32

G
γ-orthorhombic boron, 2, 121, 122, 128

H
Hole density (HD), 6, 14, 15, 18, 21, 32–35, 37, 39–46, 62
Hydrogenated borophene, 89–106, 113
Hydrogen boride (HB), v, 94–98, 102–106

I
Ir(111), 69, 70, 109, 110

M
Magnetic properties, 84, 147–149
Metallicity, 56, 73–76, 85

N
Nanotube, v, 1, 5, 7–9, 15, 31

I. Matsuda, K. Wu (eds.), *2D Boron: Boraphene, Borophene, Boronene*,
https://doi.org/10.1007/978-3-030-49999-0

O
Optical properties, 143–147

Q
Quasicrystal, 151–152

S
Self-compensation properties, 139–140
Semimetallicity, 97–102
Stuffed boron fullerenes, 29
Superconductivity, 46, 62, 79–83, 85, 150

The manufacturer's authorised representative in the EU is Springer Nature Customer Service Centre GmbH, Europaplatz 3, 69115 Heidelberg, Germany. If you have any concerns regarding our products, please contact ProductSafety@springernature.com

Printed and bound by CPI Group (UK) Ltd, Croydon, CR0 4YY
30/09/2026
02225306-0011